TANKERS
in the
SKY

The previously untold story of the huge U.S. Air Force aerial refueling fleet

CLAUDE G. LUISADA

Tankers in the Sky: The Previously untold story of the huge U.S. Air Force aerial refueling fleet

Produced and printed

by Stillwater River Publications.

Visit our website at

www.StillwaterPress.com

for more information.

First Stillwater River Publications Edition

ISBN: 978-1-968548-24-7

1 2 3 4 5 6 7 8 9 10

Cover design by Matthew St. Jean.

Published by Stillwater River Publications,

West Warwick, RI, USA.

This book is humbly dedicated to the thousands of military personnel who daily fly aerial refueling missions, past, present and future, and who have made it possible for the U.S. military to extend its influence across the entire globe.

This book is humbly dedicated to the thousands of military personnel who daily fly aerial refueling missions, past, present and future, and who have made it possible for the U.S. military to extend its influence across the entire globe.

Contents

INTRODUCTION

DEFINITION: Air-to-air refueling, also known as in-flight refueling, refers to the aerial operation whereby fuel is transferred from one aircraft in flight, known as the tanker, to another aircraft in flight known as the receiver. It is also referred to simply as *aerial refueling.*

Aerial refueling first became part of the U.S. military arsenal in the late 1940s and has grown since. It can legitimately be considered either a weapons system or, at the very least, part of a weapons system. Seen as a weapons system, the entire history of aerial refueling is worth recounting since it has played a critical role in the military operations of the U.S.

Beyond that, there are so many military aerial refueling flights every 24 hours that they constitute a significant percentage of all high-altitude flights, whether military, commercial, or business.

This, then, is the story of aerial refueling, from its experimental beginnings, through its development and growth, to its role in today's world. It is a part of aviation and military operations that is largely unknown by the public. However, it has become such an integral part of the military establishment that it deserves to be revealed to anyone who is interested in aviation and national defense.

This story is also important for two other reasons. Aerial refueling is truly a little-known facet of military aviation. Considering the impact it has had on military operations, it is certainly a subject for anyone interested in the military and military aviation. Additionally, the military's realization of the true value of aerial tankers has resulted in very large acquisition and manufacturing programs that are still ongoing. This means large expenditures of government funds. This book will, hopefully, shed light on how and why aerial refueling has become such a large component of the U.S. Air Force and why it continues to have a major impact on military planning. Aerial refueling has also played a role, although a smaller one, in the air operations of the Army, Navy, and Marines.

PROLOGUE

Loring AFB, Maine – July 16, 1965 – 02:35 A.M. EDT

The four-man crew of the KC-135A Stratotanker, radio call sign Yankee 55, is sound asleep in their alert shack, located not far from their aircraft. Suddenly, the emergency klaxon jolts everyone awake. While they do not yet know what the upcoming mission may entail, they know full well that the klaxon sounds only when a true emergency is at hand. Very quickly, the crew dons flight suits and jackets, flying boots, grabs what they call their GO BAGS, and bolts for the pickup truck waiting just outside. In no time, they arrive at Operations to be informed of the details of their mission.

The dispatcher informs them that a B-52 returning from a SAC operational mission to the edge of the Soviet Union is having a fuel problem. The fuel distribution system has malfunctioned and the B-52 is having difficulty drawing fuel from its fuel tanks. The bomber is over the far northern part of northeast Canada at 42,000 feet. Without assistance, the aircraft may have to make an emergency landing in the frozen wastes of extreme northern Canada near the Arctic Circle. Aerial refueling is the only answer.

The navigator quickly begins to plot the course to fly that will intercept the B-52 as well as the en route flight time. Meanwhile, the copilot determines both the fuel needed by the tanker to complete the mission and how much fuel the bomber will require to reach Loring AFB safely.

After hurriedly completing their flight plan, the crew is back in their pickup truck on their way to the flight line and Yankee 55. In the meantime, a fuel truck has been adding fuel to Yankee 55's fuel tanks.

With the crew aboard, the engines are fired up and the big tanker taxis toward the active runway. The control tower has already given takeoff clearance since they had been alerted to the urgent nature of the flight. Yankee 55 wheels onto the runway and, as takeoff power is applied, rapidly gains speed, lifts off, and disappears into the night.

Two hours later, some 1,000 nautical miles northeast of Loring AFB and high over northern Canada, Yankee 55 makes radio and radar contact with the disabled B-52. The two aircraft hook up, and enough fuel is transferred to keep the bomber in the air long enough for it to reach Maine. Due to the problem the B-52 is having, it is decided that Yankee 55 will escort the B-52 back to Loring AFB, periodically transferring additional fuel.

Sometime later, as both aircraft are preparing to make their descents, Yankee 55 makes a final fuel transfer to the bomber. Then both aircraft make their approaches. Yankee 55 has already told the bomber to land first as a precaution in case it needs to take on more fuel just before actually landing.

The B-52 lands safely, followed shortly by the tanker. In the operations shack, the B-52 Aircraft Commander (A/C) seeks out his counterpart in the tanker. After shaking hands, he says, "Thanks for getting us back! It would have been a very long and very cold walk back to civilization if we had to make an emergency landing up there. What do we owe you?"

Yankee 55's A/C grins and says, "How about a cup of coffee for me and my crew?"

The B-52 A/C grins back and says, "Done." The two crews walk together to the flight line cafeteria.

Another routine tanker mission is completed successfully and professionally. Without it, the B-52 would very possibly have crashed in far northern Canada and its crew would have been killed.

CHAPTER 1
LET'S TRY SOMETHING NEW

Possibly the earliest thought about aerial refueling occurred in 1917 in Russia. A young pilot named Alexander de Seversky was flying a fighter plane in formation with a bomber. The bomber was trailing a radio antenna. De Seversky, being in an open cockpit aircraft, flew close behind and below the bomber, reached up and grabbed the antenna.

This led him to think about refueling an aircraft while in flight. De Seversky emigrated to the U.S. shortly thereafter, became an aeronautical engineer, and eventually founded the respected aircraft manufacturing company that bore his name. After reaching the U.S., De Seversky patented an aerial refueling device which he then sold to the U.S. Army Air Service.

The actual origins of air refueling can be further traced back to the 1920s. After WWI, aviation found itself in a quandary. The war had spurred some advances in airplanes. But in the 1920s, there was no clear picture as to where aviation was headed. A few military officers saw aviation as a game changer if another major war were to occur. Aviation as a means of transportation was also on the minds of some aviation pioneers. But aircraft design had not advanced to the point where much was expected to happen.

In the U.S. the military services had a few courageous officers who wanted to prove the potential of aviation to their leaders. In fact, in 1921 an Army officer and pilot, Billy Mitchell, convinced military leaders to let him stage a bombing attack on a captured WWI German battleship using small bombs. The small bombs badly damaged the battleship and it eventually sank. The powerful demonstration was not lost on some Army and Navy officers. But on the whole it did not change the basic military doctrine of the U.S. Twenty years later the Japanese attack on Pearl Harbor showed the danger of ignoring those early attempts by aviation pioneers to make aviation an integral part of the military services.

In the 1920s Army fighters were of bi-wing all-wood fabric-covered construction and basically quite fragile. Moreover, the engines of that time did not have the necessary horsepower to lift large heavy aircraft into the sky. As a consequence, army aircraft were able to carry either bombs or fuel, but it was always a trade-off as to the quantity of each.

First ever air refueling - 1923

The biggest use of military airplanes in the early 1930s consisted of employing U.S. Army Air Corps fighters and pilots to fly the U.S. mail.

The net result was that a typical fighter manufactured in the mid-1920s, such as the Boeing Model 15, had a range of only 390 miles. That meant its radius of action was 195 miles or less. Clearly this short range was a powerful deterrent to the acceptance of aviation on a large scale by military leaders.

What appears to be the first recorded in-air refueling took place on June 27, 1923, between two aircraft of the U.S. Army Air Service. The two planes were Airco DH-4B biplanes using a system developed by Alexander P. de Seversky. Then on August 27-28, 1923, three of the DH-4B aircraft, two tankers and one receiver, set a record by maintaining the receiver aircraft in the air for 37 hours. The tankers had to transfer fuel nine times for a total of 687 gallons.

Also in 1923, in another instance of early air refueling, a DH-4B flew from Sumas, Washington, to San Diego, California, using four aerial refuelings en route. The two pilots who made the flight were Lt. Lowell Smith and Lt. John P. Richter.

During this period similar demonstrations took place in France and the U.K. but were not considered to be important by the military leaders of the time.

In January 1929, the newly formed U.S. Army Air Corps conducted some audacious experiments involving air-to-air refueling. These experiments were carried out in conjunction with the Boeing Company, one of the early U.S. airplane manufacturers. A group of five military pilots were chosen for the experiment with Capt. Ira Eaker in charge.

It is interesting to note that in December 1942, during WW II, Eaker, by then a Major General, assumed command of the U.S. Army Air Force Eighth Air Force. This was the organization that, with its B-17 and B-24 heavy bombers, carried out the daylight bombardment of Occupied Europe, particularly Germany.

In 1929, an early experiment involved flying a Fokker C-2, named the Question Mark, in a circular pattern over Southern California while periodically being refueled in-flight. The flight set an endurance record by remaining airborne for a truly amazing 150 hours!

Also in 1929, the U.S. Army Air Corps and the Boeing Company worked together to stage a second and even more audacious demonstration of air refueling. This time the challenge was to prove that air refueling could expedite the delivery of air mail. There were to be four tankers in the demonstration. The Army Air Corps provided two C-1 transports and Boeing two modified Model 40 aircraft. The actual mail plane was a Boeing Model 95, a new design, nicknamed the Hornet Shuttle. The Model 95 was a biplane specifically designed to carry mail and cargo. For this demonstration, the Hornet Shuttle was modified with extra fuel tanks added, a fuel receiving unit installed, and a second seat added for an additional pilot.

On Aug. 27, 1929, the Hornet Shuttle departed Oakland, California, with Ira Eaker and Bernard Thompson aboard. The aircraft was to fly non-stop to New York City with Boeing flight crews refueling the mail plane over Elko, Nevada, and Cheyenne, Wyoming, while Army Air Corps crews would refuel the plane over Omaha, Nebraska, and Cleveland, Ohio. The Hornet Shuttle made it to New York City without incident in spite of storms and fog encountered en route. In fact, the weather over Cheyenne was so bad that the refueling was conducted at an altitude of only 500 feet, an altitude that today is considered to be instrument weather.

The Hornet Shuttle reached New York City in 28 hours and 25 minutes. The pilots dropped the mail bag that originated in San Francisco over Mitchell Field and then headed back west non-stop to complete the round trip.

While refueling over Cleveland, Ohio, a problem arose. One of the four 5-gallon oil cans that were being transferred along with the fuel dislodged from the tanker and slammed into the Hornet Shuttle, puncturing the oil tank and just missing the wing main spar. Eaker immediately landed at Cleveland for repairs. Then Eaker and Thompson flew back to Mitchell Field to fly the westward leg of the round trip. Once again they took off and flew west. After 33 hours, the Hornet Shuttle reached Oakland, California, completing the round trip. Not satisfied with what they had accomplished, Eaker and Thompson headed east again. But this time, a clogged fuel line over Salt Lake City finally brought the flight to an end. The flights convinced Eaker that aerial refueling was important to the future of military aviation.

In July 1929, a group of Army Air Corps pilots led by Major Carl Spaatz, another officer who would lead in the WWII bombing of Germany, flew a plane named the *Question Mark* over Los Angeles for over 150 hours, setting an endurance record. A year later, an even more astonishing flight took place. On July 4, 1930, the brothers John, Kenneth, Albert, and Walter Hunter took off flying two Stinson SM-1 Detroiters—one a receiver and the other a tanker. Flying over Chicago, Illinois, the receiver stayed aloft for a truly amazing 553 hours and 40 minutes. That equals just over 23 days of non-stop flying!

During the 1920s, pilots in the U.K. were also demonstrating the potential of aerial refueling. Sir Alan Cobham, a Royal Flying Corps pilot in WWI, made a number of very long flights to Africa and Australia. However, refueling, in the primitive way it was done, was very dangerous, with spills always possible. In 1935, the brothers Fred and Al Key demonstrated a spill-free refueling nozzle designed by A.D. Hunter. That same year, the Key brothers, flying a Curtiss Robin monoplane, managed to exceed the record set by the Hunter brothers by nearly 100 hours, staying aloft for 27 days!

In the meantime, military pilots in France and the U.K. continued their own experiments with aerial refueling during the late 1920s. Then, in 1934, a Royal Air Force (RAF) officer, Richard Atcherley, developed a new and safer system for refueling. He had noticed the dangers of the existing fueling systems. His invention, the "crossover system," called for the tanker to trail a large hooked line that would reel in a similar dropped line from the receiver, thus allowing fuel to flow. Cobham bought the rights to this system, and by the late 1930s, had improved it to the point that it became the first practical refueling system in the world.

Once again, it was Cobham who came up with something new. His company developed a system called the *grappled-line looped-hose.* This system was an improvement on a previous system patented by David Nicolson and John Lord. In this system, the receiver trailed a steel cable that was then grappled by a line shot from the tanker. The line was hauled back to the tanker, where the receiver's cable connected to the refueling hose. The receiver would then haul back its cable, thus bringing the hose to it.

In 1934, Cobham founded a company called Flight Refueling Ltd. By 1938, this system was used to air-refuel large flying boats. In the summer of 1939, British Empire Airways flying boats successfully completed sixteen trans-Atlantic crossings carrying mail. This service was interrupted by the start of WWII in September 1939. Nevertheless, this was the first commercial application of in-air refueling. All these demonstrations did little to change the way military leaders of the time viewed air refueling. But they did greatly expand the thinking of young military pilots of the period, such as Eaker and Spaatz. Decades later, the success of those demonstrations would play a major role in the development of post-World War II military aviation. It is true that these early flights were nothing more than experimental and demonstration flights. Yet, as we shall see later in this book, they were very valuable because they gave certain U.S. military leaders a glimpse of what was possible. It was a peek into the future, and 20 years later, when the young pilots of the 1920s had become the leaders of military aviation in the 1940s, they remembered those early experimental flights and used that knowledge to good advantage. From those tiny beginnings, there would, in time, spring forth a mighty armada of aerial tankers.

CHAPTER 2
THE NEED FOR AERIAL REFUELING ARISES

The story of how, in the late 1940s, the need for aerial refueling became apparent, is long and convoluted. But it is worth following because it shows the chain of events that led to that need.

During WWII, there was no attempt to utilize aerial refueling. This was true despite the fact that lack of range seriously affected many military air operations on both sides.

Fig. 2.1 German Luftwaffe Heinkel 111 medium bomber

In the summer of 1940, the German Luftwaffe tried hard to bomb England into submission by daily mass raids on London. Its twin-engine bombers flying from French airfields were escorted by fighters. But the fighters had so little range that they could remain over England for only a few minutes. This shortcoming gave the Royal Air Force (RAF) Hurricanes and Spitfires an excellent opportunity to shoot down the German bombers. Despite the Luftwaffe's overwhelming numerical superiority, they suffered such losses that after three months it cancelled the daylight raids and switched to night raids.

In 1944, B-17 and B-24 heavy bombers of the U.S. Army Air Force's Eighth Air Force were making deep-penetration raids into Germany unescorted by fighters. This was one reason for the high rate of losses on these long-range bombing raids. Later on, when the P-51 Mustang, with its long-range, began to escort the bombers, the rate of losses dropped dramatically. In fact, U.S. bombers operating from Italy and bombing Germany, escorted by P-51 Mustang fighters for the entire mission, had far fewer losses.

Fig. 2-2 B-17 bomber escorted by A P-51 Mustang Fighter

In 1944, an attempt was made to solve the problem of sufficient range by experimenting with shuttle flights. These involved the heavy bombers carrying out their deep-penetration raids and then continuing on to airfields in North Africa, Italy, and Russia. There the bombers would remain overnight, refuel, take aboard new bomb loads and ammunition, and the next day fly back to England, using the return flights to complete additional bombing raids. For various reasons these shuttle flights were not considered particularly successful.

Refueling Attracts Notice

Shortly after the Japanese attack on Pearl Harbor, the U.S. Army Air Force began early experiments in aerial refueling. The Air Force had three concepts in mind as follows:

1. Operating B-17 Flying Fortresses from Midway Island refueled by B-24 Liberator bombers modified to act as tankers.
2. Operating B-24 Liberators from Hawaii with Navy seaplanes in support as tankers.
3. Using B-17 Flying Fortresses towing gliders full of fuel.

Initial testing using a modified British *looped-hose system* began in the summer of 1943 at Eglin Field, Florida. A B-17E was used as the receiver and a modified B-24D as the tanker. The test showed that a B-17 carrying three tons of bombs would have its range extended from 1,000 to 1,500 miles. In the end, nothing came of this initiative for two reasons. The demand for new military airplanes took up all the existing aviation manufacturing facilities, leaving none for the conversion of the B-17s and B-24s. In addition, by 1943, Boeing was beginning to roll out a new long-range heavy bomber, the B-29 Superfortress. Nevertheless, it is noteworthy that as far back as 1942, the military was becoming actively interested in the potential use of aerial refueling.

Then in 1944-45, U.S. Army Air Forces' 20th Air Force B-29s, based in the Mariana Islands, began bombing raids over the Japanese home islands. The round-trip distance was around 3,000 miles, just within the

B-29's 3,250-mile range, with little to spare. However, it was not uncommon for westbound B-29s to encounter very strong headwinds, which in turn made the round trip far more risky. This was the main reason why the U.S. decided to invade Iwo Jima, where B-29s, either damaged by enemy gunfire or short on fuel, could land. Here again, the question of range came up.

In 1947, the U.S. administration, under President Harry Truman, recognized the validity of air-minded U.S. Army leaders and created the U.S. Air Force as a separate and equal branch of the military services, with General Carl A. Spaatz as its commander. This new branch of the military was divided into a number of major commands, one of which was the Strategic Air Command (SAC). The first commander of SAC was General George Kenney.

The Coming of the Cold War

While there is no official date marking the beginning of the Cold War between the Soviet Union and the United States, January 1947 is often cited as such. After WWII, the Soviet Union embarked on a mission of

Fig. 2-3 Map showing expansion of the Soviet Union

expansionism. In 1945, it annexed Czechoslovakia. This was followed by Romania in 1946, Poland in 1947, and Yugoslavia in 1948. The Soviet Union already controlled East Germany and Austria by reason of treaty.

What was behind this need to control all of Eastern Europe? This writer believes there were two fundamental reasons. One was the greed and ambition on the part of the Soviet Union's ruler, Josef Stalin, to acquire territory for such basic items as crops to grow food, oil deposits and refineries, and skilled workers in factories, all to replenish the huge losses suffered in WWII. The other factor that preyed on Stalin's mind was fear of invasion from the West, led by the United States. This fear was largely based on the WWII invasion by Germany and the human losses of over 20 million men, women, and children. Stalin was also likely afraid of the fact that America had nuclear weapons.

The Soviet TU-4 Fleet

These fears on the part of the Soviet Union's leader led to another and more urgent problem. During the 1944-45 B-29 raids on Japan, three B-29s wound up in the hands of the Soviet Union. In July 1944 a B-29 over a target in Manchuria was attacked by fighters and also received a direct hit by flak. With one engine dead and a second engine losing oil, the aircraft commander elected to divert to Vladivostok, Siberia, rather than risking a landing on Japanese soil. The crew was eventually repatriated, but the B-29 was kept by the Soviets. Then, in November 1944, a B-29 on a night raid over the Japanese home island of Kyushu ran short of fuel. Again, the crew of this aircraft elected to land at an airport in Vladivostok, Siberia. Later in November a third B-29 landed in Siberia. While all the crews eventually made it back to the U.S., the three aircraft were never seen again.

The B-29s were evidently turned over to the Tupolev Design Bureau. There, under the guidance of Andrei Tupolev, a noted Russian aviation engineer, the B-29s were *reverse engineered.* This meant that an aircraft was taken apart very carefully, component by component, with drawings then made of each component. Once this was done, the

Soviets set up a manufacturing plant to build exact duplicates of the B-29. During the period 1947-49 a total of 847 of these duplicates, identified as TU-4s, were built.

It was this bomber force that concerned U.S. military leaders, particularly since by 1949 it was known that the Soviet Union also had the capability to manufacture nuclear weapons. While the TU-4 might not have the range to fly a round trip from the Soviet Union to the U.S., it certainly could reach American targets flying one-way polar routes on suicide missions. At the time, military planners who knew the TU-4s might not have the range for a round trip, theorized that Soviet flight crews, after dropping their bombs on U.S. territory, might fly back to where they could ditch in the ocean and be picked up by Soviet ships waiting there. In fact, the concern was so great in Washington that it spurred the U.S. to carry out a number of defensive countermeasures.

In the 1950s the U.S., at great cost, built three radar warning lines or nets. The first line was the Pinetree Line, which extended from coast to coast along the U.S.-Canadian border. The second radar line, the Mid-Canada Line, extended across central Canada. The third radar line, the Distant Early Warning Line or DEW Line, was built north of the Arctic Circle.

In addition, fleets of USAF and Navy aircraft, all carrying long-range radar sets and large radomes were deployed hundreds of miles from the Atlantic and Pacific coasts. These were known as the Atlantic and Pacific Barriers.

In addition, a USAF Auxiliary called the Ground Observer Corps (GOC) was created and staffed entirely by volunteers. This organization was patterned after the Ground Observer Corps created in England that operated in World War II. In the U.S., the GOC was created in 1950 and disbanded in 1956. At its peak it had 750,000 volunteer observers who reported by phone to 73 Filter Centers, also manned by volunteers 24 hours a day.

These very expensive measures taken by the U.S. clearly show the level of concern on the part of both civilian and military leaders. All these measures, however, were defensive in nature.

But things were about to change. In response to the very dangerous expansion of the Soviet Union, the U.S. would begin the process of creating an airborne offensive force. This process eventually showed the need for aerial refueling. As is often the case, when a national priority arises, events begin to occur at a much faster rate. This would be the case here, as described in the next chapter.

CHAPTER 3
SAC AND THE FIRST TANKERS

In March 1946, the U.S. Army Air Force created the Strategic Air Command (SAC), an organization whose main mission was to act as a deterrent to the Soviet Union's growing expansion and ambitions. Its first commander was General George Kenney, appointed in March 1946. Initially SAC was made up of the 20th Air Force, which was transferred from the Marianas Islands. From bases located there, the 20th Air Force had carried out the World War II B-29 raids on Japan. Now, as part of SAC, it brought its complement of B-29 heavy bombers to bases in the U.S.

Unfortunately Gen. Kenney was not a good fit for SAC. During his tenure as commander, discipline was lacking, the flying was more like that in a flying club, and the maintenance of the B-29s was rather poor. In fact, conditions were bad enough that the Pentagon decided an investigation was necessary. The result of that investigation was that, in October 1948, the Pentagon relieved Gen. Kenney as the commander of SAC and appointed Lt. Gen. Curtis LeMay as the new commander.

Gen. Curtis LeMay Assumes Command

LeMay had been the commander of the 20th Air Force in World War II. He was a hard-driving individual, often disliked by officers and enlisted men alike who were under his command. However, he got results and he was generally respected. It was Curtis LeMay who, in 1944-45, had led the B-29 Superfortress raids against the Japanese home islands. LeMay was probably the strongest advocate of long-range bombardment in the U.S. Air Force. In the late 1940s, various events eventually led him to embrace the need for USAF tankers. It is worth reviewing these events.

Lt. Gen. Curtis LeMay

Le May set about his job with characteristic energy. He moved SAC headquarters from Washington, D.C., to Offutt Air Force Base, outside of Omaha, Nebraska. With Cold War tensions only increasing, Le May believed he had to move quickly. Furthermore, Le May believed that an offensive force had to be created that was of such strength that it would constitute a warning and a deterrent that the Soviet Union could not possibly ignore. Unfortunately, in late 1948, Le May had only the beginning of what he considered an adequate force.

In 1948, SAC had an initial complement of 486 B-29s and 35 B-50s. The B-29s were left over from those used by the 20th Air Force in 1944-45 in its campaign against Japan. Many of them lacked proper maintenance. In fact, many of the B-29s were barely in flying condition.

The new Boeing B-50s were an upgraded version of the B-29. The B-50, equipped with more powerful engines, had a longer range than the B-29, but still nowhere near enough to reach the Soviet Union.

True, in 1946, SAC had begun receiving deliveries of the brand-new and very large Convair B-36 heavy bomber. The B-36 had the range to reach the Soviet Union via a polar flight path and return. However, the B-36 had initially been plagued by engine fires. Furthermore, Le May viewed the B-36 as an interim aircraft. What he really wanted were two Boeing jet bombers that were being developed. These were the B-47 medium bomber and the B-52 heavy bomber. Both of these aircraft really needed air refueling to operate at their full potential. Thus, Le May had plenty of reasons for wanting air refueling to develop into a full-fledged program.

The KB-29 Arrives

At this point, it would be beneficial to define the three terms relating to aircraft range.

The use of terms such as range, combat range, and ferry range may be confusing to the reader. Therefore, here are the generally accepted definitions of these terms:

Range – The total distance an aircraft can normally fly.

Combat range – The distance an aircraft can fly to a target carrying ordnance or bombs and still have enough fuel left to return to its base.

Ferry range – The one-way distance an aircraft can fly minus all ordnance, bombs, and other equipment, carrying extra fuel tanks and operating at its most efficient speed.

Now LeMay set about reorganizing SAC and acquiring new top-level officers. That done, LeMay could now turn his attention to how to best utilize the fleet of B-29s, B-50s, and B-36s that made up SAC. Of these three aircraft, only the B-36 had the necessary range to carry out the basic SAC mission of deterrence. The lack of range of SAC's fleet of other bombers forced LeMay to turn to the possibility of aerial refueling. At the time, the only practical method for aerial refueling was the previously described *grappled-line looped-hose system.* This system had been developed in England by Sir Cobham's company, FRL, beginning in the late 1930s but not fully developed until the mid-1940s because of WWII. In 1942, there had been some successful demonstrations of this system during trials at Eglin AFB, Florida. While it had been proven to work, it was not a great system. But at the time it was available and thus was chosen for use by SAC.

Refueling System

LeMay got a contract, awarded to the Hayes Aircraft Corp. of Birmingham, Alabama, for the conversion of a portion of SAC's B-29 complement of bombers to both tankers and receivers. The contract called for 92 B-29s to be converted to tankers and 74 B-29s to be converted to receivers. The conversion of all these B-29s to tankers and receivers was a complex and expensive project.

First, 92 B-29s were converted to tankers and designated as KB-29M aircraft. This conversion involved the following steps:

1. All gun turrets and associated equipment were removed.
2. A 2,300-gallon fuel tank was installed in the bomb bay. This additional tank was connected to the normal aircraft fuel system.

3. Special rendezvous equipment was installed that would aid the tanker in locating its receiver aircraft and vice versa.
4. The fueling system was installed. This included hoses, reels, winches, and fuel pumps.

Following that initial conversion, the 74 B-29s designated to become receivers were converted and identified as KB-29MR aircraft. These aircraft required the following modifications:

1. All gun turrets and their associated equipment were removed except for the tail turret.
2. A 2,300-gallon fuel tank was installed in the rear bomb bay.
3. In some cases, the left- and right-hand gun sighting blisters were removed and replaced by 10-inch-diameter windows in order to reduce drag.
4. The forward bomb bay was modified to carry an atom bomb.
5. Rendezvous equipment was added to help the bomber locate its tanker.
6. In order to receive fuel, a *refueling nozzle receptor* was installed on the lower right side of the fuselage.

The *grappled-line looped-hose* refueling system chosen by SAC for the KB-29M tankers and the KB-29MR receivers/bombers was, to say the least, cumbersome. Furthermore, the refueling could be lengthy, depending on weather conditions. Described below is how this system functioned.

Here is how this system worked. The tanker had a 450-foot cable called the *hauling line*that came out of a hole in the nozzle receptor. A winch that was remotely controlled let the hauling line out. A crewman sitting in the tanker tail turret controlled this cable. The end of the cable was equipped with a dish-shaped pan fitted with grappling hooks. To transfer fuel, the tanker had to position itself above and behind the receiver. The tanker would let out about 300 feet of *hauling line.* A 50-

pound weight attached to the hauling line made it hang almost straight down. At the same time, the receiver/bomber let out 300 feet of *hauling line*. The tanker would then cross over the receiver/bomber. The tanker's hauling line would grapple the receiver's hauling line, and the tanker nozzle could then be hauled into the receiver. After the tanker nozzle was connected to the receiver's fueling system, fuel would begin to flow. In most normal operations, the fuel was directed to the 2,300-gallon fuel tank located in the receiver's rear bomb bay.

In the summer of 1948, the USAF created two Air Refueling Squadrons: the 509th based at Walker AFB, Roswell, New Mexico, and the 43rd Air Refueling Squadron based at Davis-Monthan AFB, Tucson, Arizona. By January 1949, both squadrons had received their full complement of aircraft.

The two squadrons proved the system would work operationally, but it was definitely very unwieldy. Lengthy refueling at night was probably almost impossible. In turbulent air, a grappling line could tear loose. SAC knew it had to find a much better system. However, the KB-29M proved that aerial refueling could be done operationally in a couple of missions in the early 1950s. In 1952, 60 F-84G fighters flew non-stop from Georgia to California thanks to being refueled en route by 24 KB-29M tankers. A year later, in 1953, F-84G fighters flew from Georgia to French Morocco with three in-flight refuelings, although in that case the tankers were the new Boeing KC-97 model tankers.

Sometime during this period, the Boeing Company modified 116 KB-29 tankers to carry the first version of a tanker refueling boom.

The 43rd Air Refueling Squadron at Davis-Monthan AFB had a total complement of 25 of the KB-29M tankers. They operated these from 1948 to the end of 1953. By that date, they had been replaced by the new KC-97 tanker. Of the original complement of KB-29M tankers, 14 were transferred to other units during 1953. The remaining 11 aircraft all went into storage. By June 1954, all the tankers of the unit were KC-97 tankers. The 509th Air Refueling Squadron retired its KB-29M tankers shortly thereafter.

It is not easy to evaluate the level of success of the KB-29M tankers. The type of refueling equipment was definitely not very practical, and the B-29 airframe was far from ideal as a tanker. Yet, considering that this was the very first large-scale attempt at air refueling bomber aircraft in flight, it must be considered a real success.

By creating operational refueling squadrons, SAC accomplished two very important goals. First, as more and more refueling missions were undertaken, the crews increased their proficiency and even managed to reduce the time it took to refuel a bomber. Second, as the months passed and more missions were completed, operational units and SAC developed better and more efficient procedures. This included not only the actual refueling, but also such things as tankers being able to locate the receivers in poor weather and at night.

One might look at the entire refueling operation conducted with the KB-29M tankers as an experiment and a proving ground for the much greater level of refueling that was to develop in the near future. As such, this first experiment must be considered an unqualified success. It was that necessary first step that would lead to routine large-scale air refueling for both SAC and the Tactical Air Command (TAC). As the KB-29M tankers were being phased out, beginning in 1951, a new USAF tanker force was already being formed. The new KC-97 tankers were being delivered by Boeing, and new tanker squadrons were being formed as the entire SAC tanker force grew exponentially. The KB-29M tanker force constituted that necessary first experimental step that would, in short order, lead to routine large-scale air refueling the world over. Seen in that light, it was a small step that would shortly lead to a far larger tanker force, the first in history.

CHAPTER 4
BAPTISM OF FIRE FOR THE TANKERS

In 1950, the brand-new USAF tanker force suddenly found itself in the middle of a shooting war. On June 25, 1950, ground forces of the Communist country of North Korea invaded the nation of South Korea. U.S. President Harry Truman, together with other world leaders, condemned the invasion. A multi-nation coalition led by the U.S. was created to fight to save the nation of South Korea.

The U.S. was ill-prepared for this war. After World War II, the U.S. had greatly reduced the size of all the branches of the military. Nevertheless, troops were rushed to the battle zone to reinforce South Korean ground forces. The North Korean Army managed to force coalition forces further and further south until all that remained was a small area known as the Pusan Perimeter. Then, in September 1950, coalition troops were landed at Wonsan on the west coast of the Korean Peninsula. This landing and the subsequent military action cut off much of the North Korean Army and forced it to retreat.

When the Korean War first broke out, the USAF was ill-prepared to fight. Furthermore, its aircraft had to operate from Japan since there were no airfields available in South Korea during the first few months of the war. The B-29 bombers had the range to be ferried to Japan, but F-80A and F-84G fighters did not. To reach Japan, the fighters had to island hop, and even that was beyond their capability. External fuel drop tanks with fixed refueling probes were developed and manufactured by the U.K.'s Flight Refuelling Ltd. (FRL), the company Sir Alan Cobham had created. These were installed on the fighter aircraft, but that still did not give them the necessary safe range. These fighters were then equipped with the just-developed fueling receptacles while the KB-29M tankers were converted to the also-new Boeing refueling flying boom and identified as KB-29P tankers.

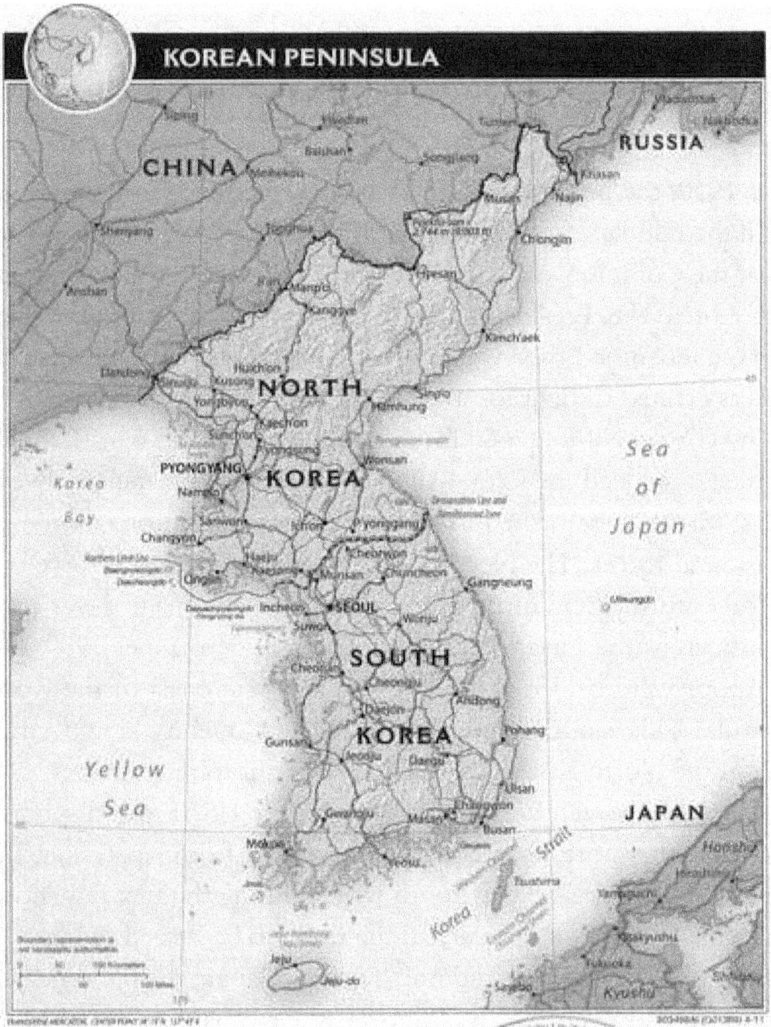

Fig. 4-1 Map of Korea

In the meantime, the Air Force also found itself facing additional new challenges halfway around the world in Europe. The need arose to ferry some F-84 fighters to West German airfields from their base in Georgia. To accomplish this movement, the jets had to make the trip with fuel stops at Bangor, Maine; Goose Bay, Labrador; Bluie West One Airfield, Greenland; Iceland; and Stornoway, the Hebrides Islands. The entire movement required a total of ten days. This was considered simply unacceptable by some Air Force officers.

In October 1950, two modified F-84E fighters flown by Colonel David Schilling and Colonel William Ritchie proved what aerial refueling could accomplish. Departing westbound from RAF Manston in the U.K., they were initially air refueled by Flight Refuelling Ltd. (FRL)-operated Avro Lincoln bomber-tankers over Scotland and Iceland and by USAF KB-29M tankers further west. They reached the U.S. nonstop in just 10 hours! Ten hours instead of 10 days! Air refueling had finally and truly proven its value!

On May 6, 1951, the first combat air refueling of fighters took place over Korea. Three RF-80A reconnaissance fighters launched from Taegu, South Korea, and equipped with the newly modified wingtip drop fuel tanks, were refueled by a KB-29 tanker offshore of North Korea. The added range enabled the F-80A aircraft to take valuable photos over North Korea. In the following two years, F-84G fighters flying missions over Korea were similarly air refueled. The U.S. Navy also began air refueling using modified AJ-2 carrier-based attack bombers.

By May 1951, the USAF B-29 bombers and F-80A and F-84G fighters operating from bases in Japan began operations against the North Korean forces. But Japan was a long distance from the battleground for these fighters. It was not until the Korean War was a year old that large-scale combat refueling missions began. These were the first large-scale combat refueling missions in history.

In the latter half of 1951, a number of B-29 bombers were shot down by North Korean MiG-15 jet fighters. As a result, the B-29s

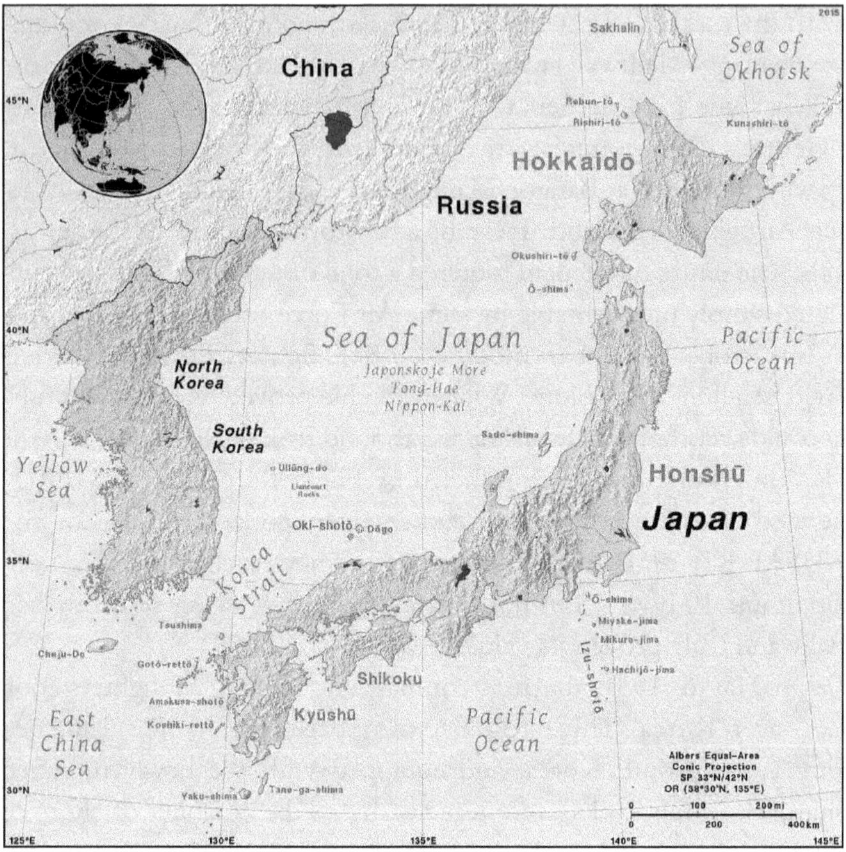

Fig. 4.2 Map of Japan and Korea

began bombing at night. It was then realized that tactical aircraft that could be air-refueled were capable of carrying out these same bombing missions.

Air refueling was even used to enable USAF C-119 troop carrier transports to fly over North Korea and parachute supplies to Marine divisions that were being overrun by Chinese troops when China suddenly entered the Korean War.

In May 1952, 12 F-84E fighter-bombers operating from a base in Japan were air-refueled and reached targets in North Korea. That same year, F-84s were able to fly nonstop from U.S. bases to Japan using air refueling.

Then, on July 4, 1952, the big test for air refueling occurred. The mission called for moving 60 F-84G fighters, an entire Fighter Wing, from their home base in Georgia to first Travis AFB, California, and then on to Japan. The mission, designated Fox Peter One, was organized by Colonel Schilling, who was previously mentioned. The fighters flew nonstop to Travis AFB, refueled by 24 KB-29P tankers over Texas. The fighters then managed to fly to Hickam AFB, Hawaii, a distance of 1,860 nautical miles, thanks to their drop tanks. From there, they island-hopped to Japan. The entire movement was safely completed in 12 days!

In October 1953, another record mission took place. It was again led by Colonel Schilling, now a Wing Commander, and it involved moving eight F-84G fighters from Georgia to French Morocco in North Africa. The mission was designated Operation Longstride. The fighters flew nonstop to their destinations with air refueling provided by KC-97 tankers operating from Bermuda and the Azores. A distance of 3,800 nautical miles was covered in just 10 and a half hours!

Air refueling was now no longer a dream of the military. It had become a reality and was about to grow further. New tanker aircraft were to improve and expand air refueling in the next few years. In the meantime, the top military leaders were becoming increasingly convinced of the absolute necessity for a large, modern fleet of aerial tankers. The events described in the next chapter show how the military leaders and Congress began to take the initial steps that would ultimately lead to a large and very active tanker force.

CHAPTER 5
THE BOEING B-50 ERA

After the end of WWII, it was not long before the U.S. found itself confronting a powerful Soviet Union in what became known as the "Cold War." In the second half of the 1940s, the Soviet Union gradually took control of all of Eastern Europe and even part of Germany and Austria. The Strategic Air Command (SAC) of the USAF, under the leadership of General Curtis LeMay, became the principal deterrent force against further Soviet expansion.

The reader may question why this chapter discusses the Boeing B-50 bomber instead of a tanker. The period from 1946 to 1955 was a transition period. SAC had a sizable striking force with the arrival of both the B-50 and the long-range Convair B-36. However, General LeMay had no doubt that aerial refueling was precisely what SAC needed as soon as possible, particularly with the soon-to-arrive Boeing B-47 jet bomber. However, LeMay was not convinced of the practicality of the aging KB-29 tanker or of the "looped hose" refueling system. In order for SAC to acquire better tankers and a more efficient refueling system, political obstacles had to be overcome. The B-50, despite its somewhat limited range and speed, helped LeMay and the Air Force overcome those obstacles. Chapters 7 and 8 discuss the results that ensued from some of the shortcomings of the various aircraft SAC had at that time. Thus, the B-50 era was very important as a transitional period that helped bring about the first refueling tanker force in history.

The Air Force and LeMay were very concerned about the relative weakness of SAC. The B-29 fleet of WWII bombers that made up much of SAC's striking force was, by this time, old and worn out, with many suffering metal fatigue and corrosion. SAC had one interim weapon in the Convair B-36 heavy bomber, which first joined SAC in 1948. But the B-36 had various problems, and what LeMay really wanted was a temporary replacement for the B-29 while awaiting the arrival of the world's first jet bomber, the Boeing B-47.

The answer was the B-50, which, to some observers, looked just like the B-29. However, there were a number of significant differences between the two aircraft.

At this point, it might be worthwhile to take a closer look at the Boeing B-50 bomber, which later became a tanker. The B-50 was very similar in appearance to its predecessor, the B-29. However, it was an airplane with many changes from the B-29 and represented a noteworthy design. Many people think the B-50 was nothing more than an improved B-29, but in reality, once Boeing finished the B-50 design, there were some very significant differences. Boeing had first worked on a B-29 upgrade for the Air Force, identified as the B-44. This design did not please the Air Force, so Boeing scrapped it. Boeing then redesigned the B-29, but with major changes.

The table below highlights some of the significant design and operating differences between the two airplanes.

Characteristics

	B-29	**B-50**
Wingspan	141’ 2”	141’ 3”
Tail Height	29’ 7”	32’ 8”
Length	99’ 0”	99’ 0”
Cruise Speed	220 MPH	244 MPH
Combat Range	1,650 miles	2,394 miles

Ceiling	31,850’	36,900’
Max Gross to Weight	133,500 lbs.	173,000 lbs.
Horsepower	2,200 HP x 4	3,500 HP x 4

The B-50 had a single large bomb bay, unlike the B-29, which had two smaller bomb bays.

The B-50D model carried a crew of 10.

Admittedly, the B-29 and the B-50 had almost identical appearances. The fuselage and wings were essentially the same. However, the engines of the B-50 were far more powerful. This resulted not so much in greater speed but rather the capability to carry a greater payload. The B-50's maximum gross take-off weight was 40,000 pounds or twenty tons greater than the B-29. To accomplish this increase in gross take-off weight required a considerable increase in the strength of the wing structure.

The major performance improvements of the B-50 over the B-29 came as a result of the new power plants. These were Pratt & Whitney R-4360-35 Wasp Major turbo-supercharged radial engines of 3,500 horsepower each. That meant each engine produced 1,300 more horsepower at take-off than the engines in the B-29.

The B-50 wing outer panels were reinforced to handle the considerable increase in the take-off weight as well as the weight of the heavier engines. The landing gear was also redesigned to allow for the additional take-off and landing weights. The tail fin and rudder were raised to give better lateral control at low speeds.

The B-50 Superfortress was manufactured by Boeing from 1947 to 1953. The first flight was in June 1947.

In 1948, SAC began receiving the first of 370 Boeing B-50 bombers. While the B-50 had greater range than the B-29, it was still not considered sufficient to reach the Soviet Union. As a result, SAC saw the B-50 as a bomber that could only be used if air-refueled. Consequently, 222 of the B-50D model were fitted with fueling receptacles for in-flight refueling by tankers using the newly developed Boeing *flying boom* (See Chapter 7).That group of aircraft was given the designation *KB-50.*

Additionally, during that period of 1948-53, the Soviet Union had generally poor radar coverage. As a result, SAC believed that the B-50 could reach and penetrate Soviet airspace with impunity and be able to reach targets via the shortest route.

Curtis LeMay saw the B-50 as a short-term interim bomber while he waited for the much higher-performing Boeing B-47 jet bomber. As a result, a few years later, the B-50 fleet began to be transferred out of SAC to be replaced by the bigger and much faster B-47 jet bomber.

The Round-the-World Flight of Lucky Lady II

LeMay wanted the Soviet Union to have no illusions or doubts as to the ability of the U.S. to reach and bomb that country. To that end, he staged a number of well-publicized long-range flights using air-refueling. In February 1949, he staged the biggest of these demonstrations.

On February 26, 1949, a B-50D Superfortress took off from Carswell AFB near Fort Worth, Texas. The aircraft was named the "Lucky Lady II" and was quite new. It carried the normal configuration of armament. However, it was fitted with an extra fuel tank in the bomb bay. The "Lucky Lady II" also carried a double crew aboard.

The "Lucky Lady II" departed Carswell AFB, Texas, and flew east, flying over Lajes Air Base in the Azores; Dhahran Airfield, Saudi Arabia; Clark Field, the Philippines; and Hickam AFB, Hawaii. On March 2, it landed back at Carswell AFB, having completed a non-stop round-the-world journey of 23,452 miles in 94 hours. The flight was refueled in the vicinity of the four bases listed above by B-29M tankers, proving beyond any doubt that long-distance bombing missions were feasible with air refueling. However, the Lucky Lady II's round-the-world flight also proved that the grappled-line looped-hose refueling system was easily affected by wind currents and that a better system had to be developed.

While these long-distance demonstration flights were probably successful in terms of showing how the U.S. could reach targets in the Soviet Union, they were really not the sort of flights that could be conducted operationally on an ongoing basis. The grappled-line looped-hose system was awkward and cumbersome and potentially dangerous in bad weather. In addition, during bad weather, it could take a long time to complete the hookup. During that time, fuel was being consumed.

Le May believed a better system was urgently needed. He went to Boeing and explained the need for a different and greatly improved refueling system to serve his bombers. Furthermore, when Le May had assumed command of SAC in October 1948, he already knew that a new weapon system, the B-47, was on the way. This weapon system would, without question, require a far better and more reliable refueling system than what he had in 1948. A new and far better system was on the way. The role of the B-50 was about to change, and it would acquire a new mission.

In the end, the true value of the B-50 turned out to be not as a bomber but rather as a tanker. The KB-50 also very clearly proved that this was not the tanker that SAC really needed.

So now we must leave SAC and investigate how another major command of the USAF, the Tactical Air Command (TAC), also found itself in need of air refueling. The fact that two major commands were requiring the availability of tankers resulted in an even stronger demand for tankers.

Conclusion

While some might say that the B-50 era at SAC was a waste of time, the truth is somewhat different. Yes, the B-50 era was a period when SAC was still attempting to find the right aircraft as a tanker. The KB-50 was another step in the process that SAC had to follow to reach its goal. It was also a step that gave SAC time to develop what it required in a true tanker force. Seen in that light, the KB-50 did the job it was handed.

CHAPTER 6
THE TACTICAL AIR COMMAND AND THE KB-50 TANKER

While SAC was worrying about the limited range of its bombers, another major command found itself in the same predicament. The USAF Tactical Air Command (TAC) was being equipped with jet fighters beginning in 1945. In the latter half of the 1940s and the first half of the 1950s, TAC was receiving deliveries of such early generations of jet fighters as the F-80, the F-84, the F-86

Fig. 6-1 Photo of Lockheed F-80A
The first USAF operational jet fighter

84, the F-86, and the F-100. These four aircraft had relatively short ranges. The F-80 range was 825 miles, the F-84 1,000 miles, the F-86 1,525 miles, and the F-100 2,000 miles. The combat range of all these fighters was somewhat less than half of the total range.

Fig. 6-2 Photo of North American F-86 –
The USAF Jet fighter that proved its superiority in the Korean War

To put these performance numbers in context, consider the following: By 1950, the Cold War was very real, and the possibility of fighter operations in Europe was a major concern. Then, in July 1950, the Korean War began with its own unique requirements for fighter operations. In both cases, a greater combat range would have been greatly beneficial. The other problem was the ferry ranges of these fighters so that they could reach their operational bases, either in Japan/Korea or Europe, from the Continental United States (CONUS).

A vivid demonstration of how difficult it was to ferry those early-generation fighters over long distances took place in the summer of 1948. A flight of 16 F-80 Shooting Star fighters was to be ferried from

Selfridge AFB, Michigan, near Detroit, to a Royal Air Force Base in England. To accomplish this mission, the F-80s had to land and refuel at Bangor, Maine; Goose Bay, Labrador; Bluie West 1 Air Base, Greenland; Meeks Field, Iceland; and Stornoway, in the Hebrides Islands. It took the flight 10 days to complete the redeployment, delayed in part by weather. In an emergency situation, this was considered an unacceptable method of ferrying fighters to distant areas.

Refueling of fighters proved to be a different matter than the refueling of bombers. So TAC opted to equip itself with tankers that used the*hose-and-drogue* system. The big question facing TAC was what aircraft to use as a tanker.

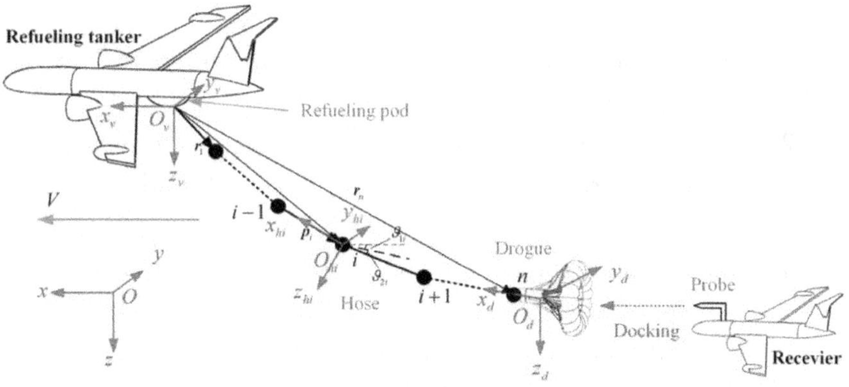

Fig. 6-3 Diagram of hose and drogue refueling system

As described in Chapter 5, SAC had acquired 370 Boeing B-50 bombers to replace its aging B-29s. As previously mentioned, the B-50s were considered to be interim strategic bombers that were fill-ins until the B-47 jet bomber was ready. By 1953 SAC had received deliveries of long-range Convair B-36 bombers and shortly thereafter Boeing B-47 jet bombers, which would initially be refueled by the new Boeing KC-97 tankers (See Chap. 8). As a result, the B-50s became available as an initial tanker for TAC's fighters.

When it turned out that SAC no longer needed the B-50s as bombers, *134* of the B-50A, B, and J models were identified as being available for conversion *to* tankers for TAC. The Air Force developed a conversion program for the aircraft to allow it to refuel fighters, and 112 were eventually converted to KB-50 tankers.

The B-50 Conversion to Tankers

TAC planned to use the KB-50 to refuel its fighters. To that end, as part of the conversion, KB-50s were equipped with *hose-and-drogue refueling systems.* This system employed a flexible trailing hose at the end of which was a cone-shaped drogue, sometimes called the "basket."

The conversion was rather complex and was carried out by the Hayes Aircraft Corporation in conjunction with the Boeing Co.

This conversion involved the following steps:

1. Fuselage lengthened by 6 feet to accommodate fueling crew-men on either side
2. Outer wing panels reinforced due to weight of refueling pods
3. Installation of hose-and-drogue fueling pods under each wing-tip
4. Installation of a third hose-and-drogue fueling line underneath the rear fuselage
5. Installation of fuel tanks in the bomb bay
6. Installation of A-12B-1 fueling drums in rear fuselage
7. Refueling operators' control stations added aft of pressure bulkhead together with observation blisters
8. Tail fin raised for better directional control
9. Installation of two J-47 5,200-pound-thrust jet engines, one underneath each wing, replacing auxiliary fuel tanks.

The addition of the two jet engines made a tremendous difference when refueling jet fighters. Now the KB-50 could refuel fighters at a speed of over 400 miles per hour at 17,000 feet. This speed was well above the stalling speed of the TAC fighters. Furthermore, the two jet engines could be used on takeoffs, which in turn led to a heavier maximum takeoff weight and to shorter takeoffs.

The ability to use three fueling systems simultaneously required a more complex arrangement of the fuel tanks and the connecting internal plumbing than the single-point system used by SAC. The great advantage of the three-point system was that three fighter aircraft could be refueled simultaneously.

A fueling system was attached below the belly of the aircraft at the very rear. Then a refueling system was attached under each wing in a pod. The KB-50 had to carry two kinds of fuel. In its wings, it carried high-octane aviation fuel for its Pratt & Whitney reciprocating engines. Inside the fuselage, it carried jet fuel for refueling jet fighter aircraft, as well as for operating the KB-50's jet engines under the wings.

On a typical mission, the KB-50 carried 30,550 pounds of jet fuel that could be transferred to jet fighters. On a mission where a maximum amount of jet fuel would be required, it carried 39,429 pounds of fuel. On a maximum-range mission, the amount of jet fuel carried decreased to 22,800 pounds.

The normal KB-50 crew numbered 8.

The other part of the total refueling system was built into the receiver aircraft—in TAC's case, its fighters. For fighter aircraft, this meant installing something called a probe. The probe is a rigid piece of pipe that extends from either the nose or a wingtip of the receiver, with a refueling valve at its outer end. The pilot of the receiver aircraft has to slowly edge up to the drogue or basket and then firmly insert the probe into the center of the dish-shaped basket attached to the end of the cable. Once the probe is firmly inserted in the drogue, fuel begins to flow automatically.

The first KB-50 tanker entered the TAC inventory in January 1956. The KB-50s were delivered for two years, gradually replacing TAC's KB-29 tankers. By December 1957, all of TAC's fighter squadrons had their full complement of the new tankers.

The fleet of KB-50 tankers brought TAC into the refueling community and helped solve a serious problem, which was how to enable it to project its tactical power overseas.

Unfortunately, in October 1964, with KB-50 tankers deployed to Thailand to refuel fighters during the Vietnam War, disaster struck. A KB-50 tanker crashed on takeoff, killing all the crew members. The subsequent investigation revealed that extensive corrosion in both wings had caused them to snap off during takeoff. When the rest of TAC's KB-50 fleet was inspected for wing corrosion, the same problem was discovered in many of the remaining tankers. As a result, TAC immediately grounded the entire KB-50 fleet of 370 tankers, fearing more crashes. This, of course, left TAC in an awkward position operationally. The solution arrived at was for the SAC KC-135 tanker fleet, which had recently joined SAC, to do double duty and refuel the TAC fighters, provided that SAC retained control of all its tanker fleet.

Conclusion

The KB-50 tanker was, admittedly, a short-term interim solution in the continuing development of aerial refueling. Nevertheless, it was an important step and should be viewed as just that. If it had not been for the problem of wing corrosion, the KB-50 would probably have had a longer operational life. Meanwhile, waiting in the wings were new developments in the constantly growing field of aerial refueling. Furthermore, the arrival of the first jet bomber, the Boeing B-47, simply added even more urgency to the further development and growth of the tanker fleet. But first, we must examine two events that led to the true tanker fleet.

CHAPTER 7

THE B-47 BOMBER AND THE FLYING BOOM

By mid-1949, not long after he assumed command of SAC, General Curtis LeMay was confronted with three realities. These were:

1. An aging bomber fleet
2. An equally aging tanker fleet
3. A refueling system that was not very practical.

Somewhat surprisingly, the answers to all three problems lay with the Boeing Company.

SAC had already upgraded its bomber fleet with the introduction of the Boeing B-50 in 1948. A total of 370 of these bombers were manufactured. But the B-50, while slightly faster than the B-29 and with a longer range, was still not what General LeMay considered a satisfactory solution. The B-50 was merely a short-term interim solution. SAC was also receiving deliveries of the very long-range Convair B-36 bomber from 1946 to 1954. The B-36 had been designed at the end of WW II. A total of 384 were built for SAC. However, the B-36 had a number of operational problems, and it was not long before it was being called "obsolescent" by many in the Air Force. The B-36 fleet was retired in 1959 after a relatively short service life. By then

SAC had already received a quantity of Boeing B-52 long-range heavy bombers.

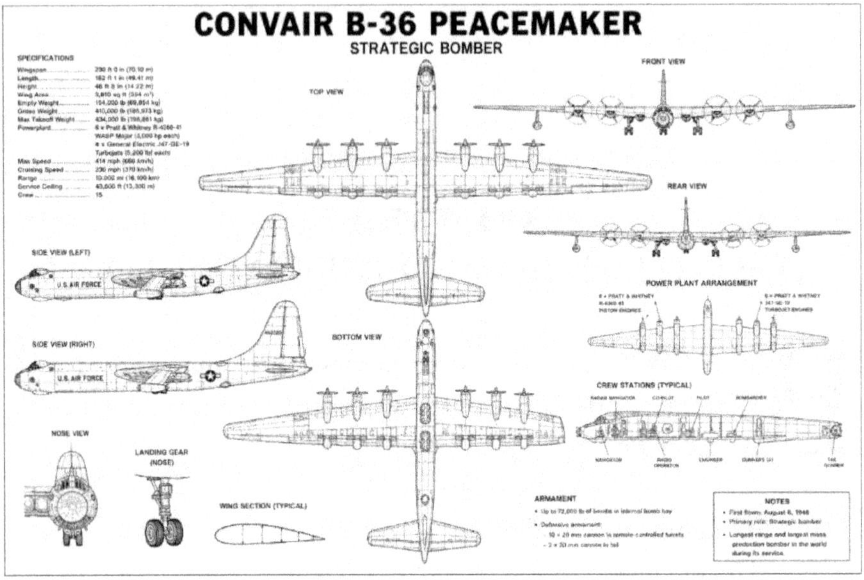

Fig. 7-1 The Convair B-36 Long-Range Bomber

The Jet Age Arrives

In mid-1951 SAC began accepting deliveries of a swept-wing jet bomber, the Boeing B-47. This aircraft represented a huge step forward in terms of speed, but it still lacked the necessary range SAC needed. The B-47 had six (6) jet engines, a three-man crew, a cruising speed of 550 mph, and a range of just over 2,000 miles. In spite of its deceptive appearance, the B-47 was a big airplane with a maximum take-off weight of 230,000 pounds. Boeing built 2,042 of these jets, beginning in mid-1951, all for SAC.

The B-47 represented a game changer for SAC. This airplane gave SAC the ability to bomb accurately and at high speed from a much higher altitude. The one thing the B-47 lacked was the necessary range to reach the Soviet Union. This, of course, meant that for many of their missions the B-47s required aerial refueling. During the 1950s SAC was also receiving deliveries of the Boeing B-52 heavy bomber. From 1952 to 1962 Boeing built 744 of the B-52s. While the B-52 had a much longer range than the B-47, it still required aerial refueling on many of its missions. This was particularly true when the B-52s were flying missions to the very edge of Soviet airspace and then had to orbit there.

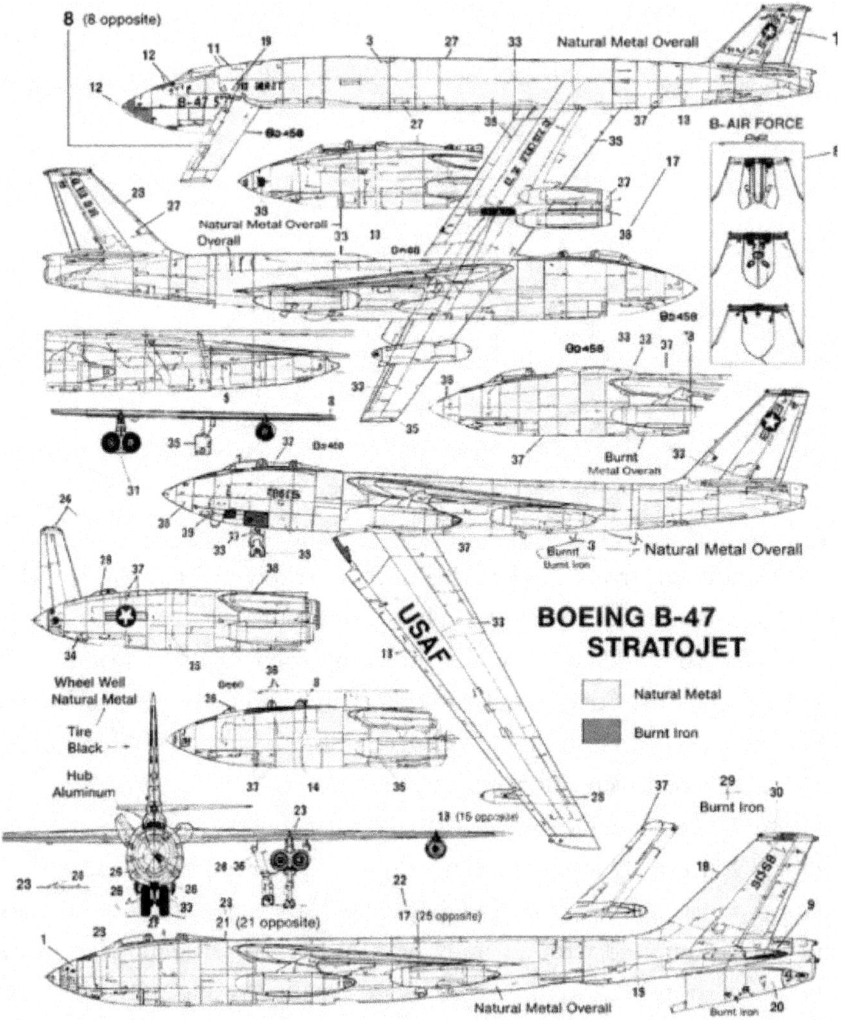

Fig. 7-2 The Boeing B-47 Stratojet

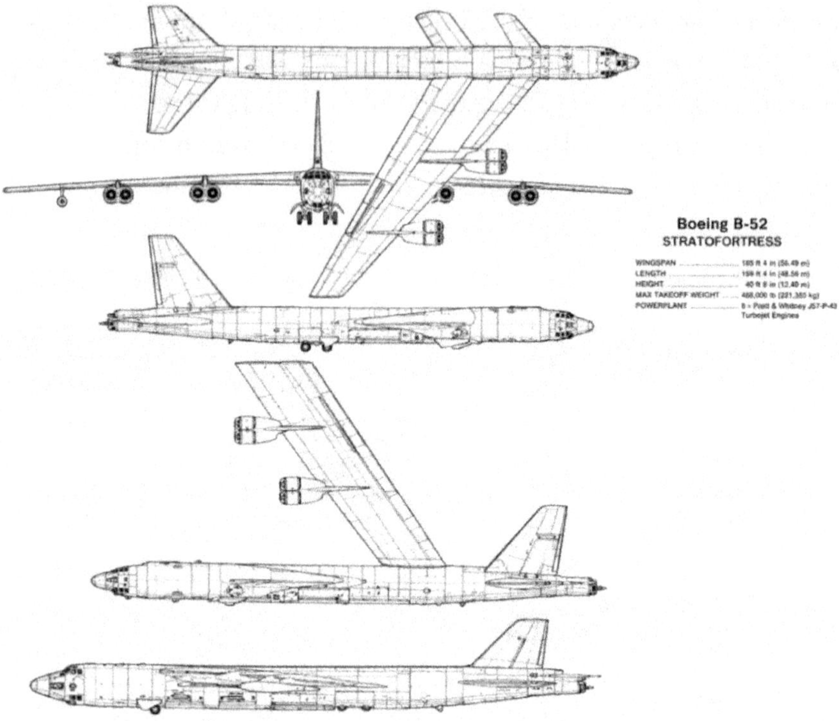

Fig. 7-3 The Boeing B-52 Stratofortress

In addition, from 1957 to 1960, SAC received 116 Convair B-58 Hustler medium jet bombers. These were able to fly at supersonic speeds for short distances while dropping their loads of bombs. The B-58 fleet also needed aerial refueling since their range was limited. This probably explains why the entire B-58 fleet was retired in January 1970 after a relatively short operational life.

Thus, the reader can see SAC's critical need for a tanker in the 1950s, even if only as an interim solution. The temporary solution lay in the KC-97, as described in Chapter 8. Any new tanker model also had to be equipped with a totally new system of refueling these three types of jet bombers. Without both a new tanker model and a far better system of refueling, the performance of the above-mentioned bombers would be severely limited.

Luckily, a temporary solution for the lack of a better tanker was at hand: the Boeing KC-97. (See Chapter 8).

The Flying Boom

The need for a replacement of the *grappled-line looped-hose* refueling system had been foreseen by a number of USAF officials years earlier. In fact, in 1946 Gen. Carl Spaatz and USAF planners had determined that a better system was a priority. Accordingly, the Air Materiel Command (AMC)of the Air Force was directed in 1948 to issue a contract to Boeing for the development of a far more viable refueling system.

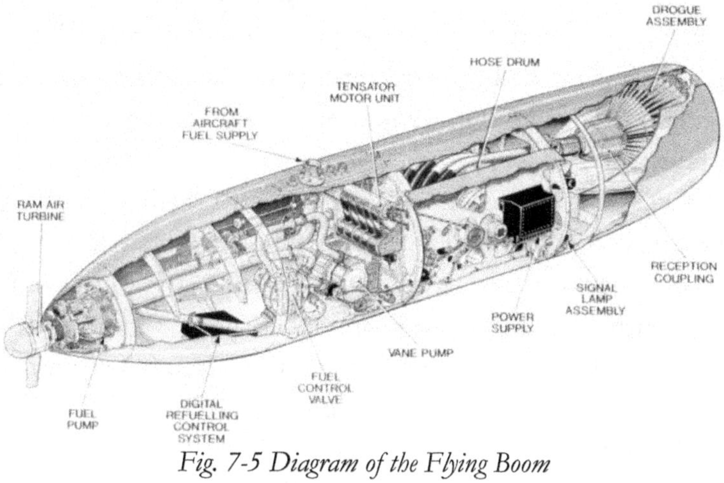

Fig. 7-5 Diagram of the Flying Boom

Boeing began work on this project and developed a new and remarkable solution called the *rigid flying boom.* The flying boom is a two-part rigid length of telescoping pipe. The larger diameter pipe is attached at one end to the tanker belly and swings down. A second pipe of smaller diameter fits inside the larger pipe. This smaller diameter pipe can be remotely extended and retracted. Near the end of the smaller pipe, a V-tail is attached, enabling the boom operator in the tanker to control the boom's motion up, down, and sideways. At the very end of the boom is a fueling valve that opens automatically. When not in use, the boom is retracted and swings up to lie flat against the belly of the tanker.

During fueling operations, a boom operator lies prone in the belly of the tanker and looks down through a small window as the receiver approaches. Once the receiver is in position below and behind the tanker, the boom operator swings the boom down and extends its inner section. Then, using controls, he "flies" the boom until its end is very close to the receiver's fueling receptacle. Then, by further extending the boom's inner section, the operator inserts the end of the boom into the receiver's fueling receptacle. Once the boom is firmly locked into the receptacle, a valve opens automatically, and fuel begins to flow into the receiver. The receiver's receptacle is usually either in the nose of the receiving aircraft or in the roof just behind the flight deck. When fueling is complete, the boom operator remotely disengages the boom, swings it up, retracts the inner section, and then stows it against the tanker belly.

To successfully transfer fuel, the pilots of both aircraft must fly their aircraft very carefully. Tanker pilots must maintain level flight and a constant speed without deviation. Receiver pilots must continually adjust for air turbulence and remain in virtually the same position below and behind the tanker. As the receiver aircraft takes on fuel, it becomes progressively heavier, and power must be gradually added for it to remain in the correct position.

There is no question that Boeing's flying boom was truly a game changer. While refueling with the flying boom still required precise flying on the part of both the tanker and the receiver, it represented a huge improvement over previous systems. Making the necessary connection between tanker and receiver could now be completed far more quickly and, in turbulent weather, had a far greater chance of being completed successfully. Furthermore, the two boom winglets gave the boom operator some real control over the operation, which was not available before. With practice, the flight crews and the boom operators became very proficient at this operation and over the years

hundreds of thousands of refueling missions were successfully completed without incident, proving the value of Boeing's flying boom. Basically, the flying boom enabled air refueling to eventually become far more routine.

Conclusion

The arrival of both the B-47 and the flying boom meant that it was now a priority to create a fleet of true aerial tankers. With the initial delivery of both the B-47 medium jet bomber and, shortly thereafter, the B-52 heavy bomber, the arrival of a true aerial tanker came not a moment too soon. The next chapter describes the first true aerial tanker and its creation.

CHAPTER 8

THE KC-97 AND REFUELING BECOMES ROUTINE

In the late stages of WWII, the Air Force had requested a new design for a large cargo aircraft and Boeing produced what the Air Force was hoping to acquire. It was identified at Boeing as the Model 367 and by the Air Force as the C-97.

The Whale Appears

In November 1944, the prototype of the C-97 transport first made its debut. The history of this airplane is quite interesting. In the mid-1940s, Boeing had developed a long-range transport based in part on the B-29 and identified as the Boeing Model 367. This became the USAF C-97. The wing, tail, and engines were basically those of the B-29. However, by the time the first production C-97 was being built, the B-50 was also under construction. As a result, the C-97 acquired tail surfaces and engines similar to the B-50. The engines, in particular, were now considerably more powerful than those of the B-29, and thus enabled the C-97 to have a higher maximum gross take-off weight.

The fuselage was totally different from that of either the B-29 or the B-50. What the Boeing design engineers had done was to take the B-50 fuselage, shave off the top, and then add a second and larger diameter fuselage on top. This resulted in a double-lobe design with a bigger upper lobe and a smaller lower lobe. In cross-section, the fuselage looked like an inverted figure eight, with a larger upper circle or lobe and a smaller lower circle or lobe. There was also a new and larger flight deck as part of the Model 367.

Fig. 8-1 Boeing C-97 prototype side view

Shortly after selling the C-97 concept to the Air Force, Boeing came up with the idea of using the Model 367 as a commercial transport. The commercial version, identified as the Model 377, was called the Stratocruiser and was marketed to the airlines as a long-range luxury passenger transport. This author flew in one in 1952 from New York to Rome, Italy, and it was definitely a luxury aircraft, with a circular staircase leading to a lower deck lounge. Unfortunately, the Stratocruiser did not sell at all well. Douglas and Lockheed had basically cornered the commercial transport market at that point. As a result, only 56 Stratocruisers were sold, nowhere near enough for Boeing to show a profit on that model.

However, when it came to the C-97, Boeing found considerable interest on the part of both the Pentagon and SAC. The Pentagon wanted a cargo aircraft capable of carrying either a large number of troops or oversized military items, such as rucksacks or even small field artillery pieces, and also serving as an airlifter. What is an airlifter? Basically, it is an aircraft able to carry a variety of military items or alternatively large amounts of humanitarian supplies.

SAC was looking for a tanker which, together with the flying boom, could refuel the B-47 jet bombers that were being built, as well as the soon-to-arrive B-52 and later the B-58 bombers. The first C-97 made its initial flight in November 1944. Eventually, the Air Force realized the C-97 could also be used as a tanker, and as such it became known as the KC-97.

Fig. 8-2 KC-97 fuselage cross-section: Note the upper and lower lobes which form a figure 8.

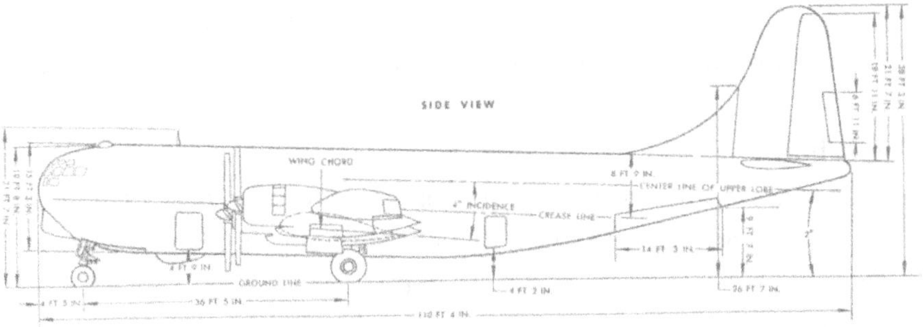

Fig. 8-3 KC-97 side view

It is interesting to note how much of the KC-97 airframe was based in part on the Boeing B-50 Superfortress, which was first delivered to SAC in 1948. The B-50, in turn, was based on the B-29. The B-50 maintained the same basic fuselage and wings of the B-29. The rudder of the B-50 was taller and the reciprocating engines were considerably more powerful, resulting in a faster cruise speed and a higher maximum take-off weight. Note that the maximum gross take-off weight of the KC-97 was 175,000 pounds.

To better visualize how the B-29 grew into the B-50 and then the KC-97, the table below compares some of the basic data among the three aircraft models.

	B-29	**B-50D**	**KC-97L**
Wingspan	141' 3"	141' 3"	141' 3"
Tail Height	27' 9"	32' 8"	38' 3"
Length	99' 0"	99' 0"	117' 5"
Mass Gross to Weight	133,500 lbs.	173,000 lbs.	175,000 lbs.
Performance Cruise Speed	220 MPH	244 MPH	220 MPH
Ceiling	31,850'	36,000'	35,000'
Total T/O H/P	2,200 hp x 4	3,500 hp x 4	3,500 hp x 4
Range	3,250 miles	2,394 miles	2,300 miles

The KC-97L tanker could carry 15,000 gallons of jet fuel for refueling and 9,000 gallons of high-octane fuel for its reciprocating engines.

The Whale Appears

It was the double-lobe fuselage, together with the higher tail fin and large flight deck, that gave the C-97 an appearance remarkably like that of a whale. In the Air Force, it was often referred to as "The Whale."

The KC-97 was a big airplane for its time. With its bigger upper lobe and slightly smaller lower lobe, it was an aircraft that had a very large interior volume. The KC-97 was truly an impressive aircraft and stood out when sitting on the tarmac alongside a mix of other aircraft. A whole row of these tankers lined up at an airfield was a unique sight.

The KC-97 had no choice but to carry two types of fuel. In its wing tanks, it carried high-octane aviation fuel for its Pratt & Whitney reciprocating engines. In the lower lobe of the fuselage were fuel tanks that carried the JP-4 jet fuel that was off-loaded when refueling a B-47 jet bomber.

Fig. 8-5. KC-97 Tanker refueling a B-47 Bomber

The upper and larger lobe could be used to carry either oversize cargo items or 96 fully equipped troops seated along the sides of the fuselage in long, fold-back canvas seats.

The flight deck of the KC-97 was very large and could accommodate four (4) crew members: aircraft commander, co-pilot, navigator, and radio operator. In addition, the KC-97 carried a flight engineer and a boom operator for a total of six (6) crew members.

The flight deck had a large number of windshields, giving the two pilots excellent visibility of the surrounding airspace.

The KC-97 Power Plants

The KC-97 was powered by four (4) Pratt & Whitney R-4360 radial engines. The R-4360 was a 28-cylinder, four-row, air-cooled radial engine. Each row had 7 cylinders. An R-4360 engine weighed a hefty 3,720 pounds and could generate a maximum of 3,500 horsepower at takeoff if wet (with water injection) for a maximum of 5 minutes. If the takeoffs were dry, the power output was 3,250 hp. METO (maximum except takeoff) power was 2,550 hp and could be used for more extended periods.

The propellers were Hamilton Standard 34G60 four-blade, solid-aluminum, constant-speed, fully-feathering units with reversing and pitch-lock-over-speed protection.

The reader might be interested in the fact that the Stratocruiser commercial transport version operated by Pan American Airways had hollow propeller blades, which gave serious problems. The individual blades had a tendency to separate from the aircraft while in flight. This happened in a number of cases, resulting in crashes.

A later model, the KC-97L, which was assigned to Air National Guard units, was equipped with an upgraded R-4360 engine that produced 3,800 hp at take-off. The KC-97L also had two General Electric J47 turbojet engines, with one engine housed in a single pod under each wing. These produced a thrust of 3,790 lb each. These jets were sometimes used at take-off if ambient temperatures were very high. More importantly, the jets were used during refueling of TAC fighters to increase the speed of the KC-97L so that the tanker's speed would be higher than the fighter's stalling speed. The KC-97L model could reach a top speed of 440 mph when the J47 turbojets were activated.

Fig. 8-6 KC-97L equipped with GE J47 jet engines refueling F-102 fighters

The KC-97 was actually not an ideal tanker for the B-47. This was because the KC-97 was too slow during refueling operations since the B-47 had a relatively high stall speed. During a refueling operation, the B-47's stall speed would increase as the bomber took on fuel and got heavier. Part of the solution to this problem was for both the bomber and the tanker to enter into a shallow descent, which increased the speed of both aircraft, making it higher than the B-47's stall speed. The Air Force called this "tobogganing."

The R-4360 radial engines of the KC-97 were also used on many other aircraft of the Air Force. Among these were the following aircraft models:

Fairchild C-119 Flying Boxcar
Douglas C-124 Globemaster II
Convair B-36 bomber
Boeing B-50 Superfortress
Boeing C-97 Stratofreighter
Boeing B-377 Stratocruiser commercial transport
Pratt & Whitney produced a grand total of 18,697 of these engines from 1944 to 1958.

This writer flew in Fairchild C-119 Flying Boxcar aircraft a number of times, powered by these R-4360 engines. It was interesting to see how the engine start differed from some other engines. At ignition, the four-bladed propeller would begin to slowly rotate and then gradually pick up speed. Engine noise rose as cylinders began to fire while large plumes of gray smoke were emitted from the engine exhausts. At full power, the R-4360 made a deep-throated bass sound.

The USAF ordered a total of 811 of the KC-97 tankers plus 77 of the C-97 freighter version. This grand total of 888 aircraft meant that the Model 367 constituted a significant financial success for Boeing and more than made up for the financial losses incurred by the commercial version, the Stratocruiser. The reader might ask why SAC wanted over 800 KC-97 tankers while it had less than half that number of operational bombers. The answer is that as Boeing began to deliver B-47 medium jet bombers, it was simultaneously delivering the first of the KC-97 tankers. In fact, with over 2,000 B-47 bombers eventually delivered, the KC-97s were kept very busy.

Boeing began deliveries of both the B-47 and the KC-97 in the summer of 1951. With the introduction of the B-47 jet bomber, the KC-97 tanker, and the "flying boom," SAC finally had the three items it needed to become a true intercontinental striking force.

The first KC-97 was delivered in July 1951 and the last in June 1956. However, even as the last of the KC-97s was being rolled out of the Boeing plant, SAC was already phasing out some of the KC-97s in favor of the new Boeing KC-135 jet tanker. Some of the KC-97s were eventually transferred to the Tactical Air Command (TAC) and were operated in Air National Guard units. The last KC-97 was retired in 1978.

The C-97/KC-97 aircraft were manufactured as various models over their production life from 1947 to 1958, as shown below.

1. C-97 – prototype – 3 built
2. YC-97 – cargo transport – 6 built
3. YC-97A – troop carrier – 3 built
4. YC-97B – with 80 airliner seats – 1 built, later redesignated as VC-97D
5. C-97A – transport – 50 built
6. KC-97A – 3 C-97As converted to tankers temporarily and used as testbeds for the flying boom
7. C-97C – MEDEVAC conversions of C-97As – 14 converted
8. VC-97D – staff transport conversions of 1 YC-97A, 2 C-97A, 1 YC-97A
9. C-97E – KC-97E converted to transports
10. KC-97E – tankers – 60 built
11. C-97F – KC-97F converted to transports
12. KC-97F – tankers with 3,800 hp Wasp Major engines – 169 built
13. C-97G – 135 KC-97G converted to transports
14. EC-97G – KC-97G converted to ELINT for CIA – 3
15. GKC-97G – 5 used as ground trainers
16. JKC-97G – testbed for under-wing jet engine installation – later identified as KC-97L
17. HC-97G – 22 converted to search and rescue variant
18. KC-97G – dual-role aircraft able to act as tankers and cargo aircraft – had under-wing fuel tanks – 592 built
19. KC-97G (troop transport conversion) – 27 converted to troop transports
20. KC-97H – 1 KC-97F converted to hose-and-drogue

21. YC-97J – 2 KC-97G converted with Pratt & Whitney YT34P-5 turboprops replacing reciprocating engines
22. C-97K – KC-97G aircraft converted to troop transports
23. KC-97L – 81 modified with under-wing J-47 jet engines and transferred to Air National Guard refueling squadrons.

As can be seen from the above list, there were many variations and conversions. But the bulk of the KC-97s were the 60 KC-97E aircraft, the 169 KC-97F aircraft, and the 592 KC-97G aircraft, for a grand total of 821 tankers.

The KC-97 Goes to Tactical Air Command

In 1956 SAC began replacing its KC-97 tankers with the brand new Boeing KC-135 jet tankers. However, the productive life of the "Whale" was far from over. The Air Force converted 81 of the KC-97E/F models to KC-97L tankers which were designed to refuel the fighters of the Tactical Air Command (TAC). This conversion required a number of fundamental changes. TAC took J47 jet engines from de-commissioned KB-50 tankers and added them to what now became the KC-97L aircraft. The J47 jet engines developed 3,800 pounds of thrust. The use of the jet engines during refueling increased the tanker's speed and made it far easier for the pilots of the receivers. Then pods were added to the wing tips so that the KC-97L could now refuel three jet fighters at one time. The interior plumbing had to be altered to accommodate the additional two fueling pods. The 81 KC-97L tankers were transferred from SAC to TAC Air National Guard units and there performed their function of refueling TAC fighters.

Flying the KC-97 Simulator

In 1963, as a volunteer and an officer in the Civil Air Patrol (CAP), the USAF Auxiliary, this writer was an escort of a small group of CAP Cadets that attended a Summer Special Activity in Oklahoma City, Oklahoma. The activity was at the FAA Aeronautical Center located near

Will Rogers Airport, but we were all housed at Tinker AFB. One day we remained at Tinker AFB and all the cadets were permitted to take turns flying a KC-97 cockpit simulator located there. After the cadets took their turn the three adult escorts also got to fly the simulator with this writer last.

The instructor controlling the simulator decided to have a little fun with me. He set the simulator up as if it were flying through a big summer thunderstorm! When the simulator was turned on there was thunder, lightning, rain pouring over the windshield and severe turbulence rattling the simulator! I did the best I could to fly the airplane but I'm sure we were all over the sky and moving hundreds of feet vertically! Everyone was laughing, especially the Air Force instructor, and so was I! It was an interesting and actually enjoyable experience.

Conclusion

It is fair to say that with the introduction of the KC-97 as a tanker to refuel its bombers, SAC had now truly reached the big time. Although the KC-97 was seen by many as just another interim aircraft, it nevertheless played a critical role during its years of operation. It enabled SAC to deploy its B-47 medium jet bombers and B-52 heavy jet bombers over a much greater area. The KC-97, together with the flying boom, also created many of the operational guidelines that are still in use today. The "Whale" did the job it was supposed to do and did it very well. It was the first true tanker and furthermore provided the link between the early experimental tankers and the jet tankers yet to appear.

CHAPTER 9

KC-97 OPERATIONS

The U.S. Air Force began operating the new KC-97 tankers in 1951. As the first of the 811 KC-97 tankers came off the Boeing assembly line, deliveries began to various SAC active-duty units. The first Wing to receive KC-97 tankers in 1951 was the 306th Bombardment Wing (Medium) located at MacDill AFB near Tampa, Florida. The second unit to receive delivery was the 22nd Bombardment Wing (Medium), located at March AFB, near Riverside, California, in 1952, while the third unit was the 91st Bombardment Wing (Medium), located at Glasgow AFB near Glasgow, Montana, also in 1952. Each of these Wings had four (4) refueling squadrons. Together, these three Wings helped work out the operating procedures and schedules for all the KC-97 squadrons that would eventually be equipped with KC-97 tankers.

In the first years of KC-97 deliveries, all the Bombardment Wings were equipped with Boeing B-47 medium jet bombers. Beginning in 1955, KC-97 tankers began refueling the first Boeing B-52 heavy bombers.

The KC-97 tankers flew all over the world, refueling both the B-47 and the B-52. When these tankers were transferred to Tactical Air Command Air National Guard units, they continued to fly missions as an integral part of the active-duty U.S. Air Force.

The role the KC-97 played as an interim tanker was critically important. The reason is that B-47 medium jet bombers had a combat range of just over 2,000 miles. This was nowhere near enough for B-47s to get close to the boundaries of the Soviet Union, which, during this period, was seen as America's main adversary. With the KC-97s able to not only refuel the B-47s on their outbound leg but also to refuel them on their return leg, the B-47 fleet now became a far more dangerous asset in the Cold War.

When B-47 Bomber Wings deployed to overseas bases, the KC-97 tanker squadrons attached to those Wings deployed with them. In other words, the tankers were an integral part of the Bomber Wings.

Refueling could, at times, be both difficult and dangerous. There were a number of reasons for this. The KC-97, when fully loaded, had a ceiling of 30,000 feet. This did not put the aircraft above all weather disturbances that might be encountered. Refueling in rough air can quickly become difficult and even dangerous. The flying boom can automatically disconnect if it is stressed beyond a certain point. Every time this happens and the receiver aircraft has to reconnect, it lengthens the total refueling time.

While the service ceiling of the KC-97 was 30,000 feet, the actual altitudes at which refueling took place could vary greatly. Weather was certainly a determining factor. Turbulence has the potential for making refueling considerably more difficult. Pilots call turbulence "chop". Light chop is usually not a problem, but moderate or severe chop can present a real challenge. The two pilots of the tanker are required to fly a smooth flight path, maintaining a straight course and an altitude that does not vary. Chop can make that difficult or even impossible.

Another issue was that the KC-97's top speed was around 375 knots, a speed that at 30,000 feet was not much higher than the stall speed of the B-47s being refueled. The partial answer to this dilemma was for both aircraft to enter into a shallow descent, thus increasing the speed of the KC-97. Another partial answer was to carry out the refueling at much lower altitudes where the B-47's stall speed was higher. Thus, weather permitting, fuel transfer from a KC-97 to a B-47 might be no higher than 15,000 feet. It should also be kept in mind that while the KC-97 ceiling was listed as 30,000 feet, it was difficult for the tanker to reach that altitude initially if at maximum gross take-off weight. But after a few hours of cruising, with some of the avgas consumed, it was more doable.

When refueling of the B-52 bombers began, the same descent profile was used. In addition, the B-52s would lower both flaps and the rear main landing gear in order to slow down their airspeed sufficiently.

In addition to the typical refueling flights, some KC-97 tankers were assigned special missions. KC-97s operating from Thule AFB in northern Greenland flew in support of B-47 reconnaissance flights in the Arctic.

The Cuban Missile Crisis

In October 1962, SAC U-2 and F-101 reconnaissance flights over Cuba took high-altitude photos that revealed the unmistakable signs of Soviet launching sites being built for medium-range nuclear missiles. President John F. Kennedy considered these sites to be extremely dangerous, not only for the U.S., but also for parts of Canada and Central America. Accordingly, SAC was placed on a war footing in order to be ready to retaliate against the Soviet Union. For 13 days, from October 16 through Oct. 29, the world held its breath, waiting to see if a nuclear war between the Soviet Union and the U.S. would explode with catastrophic results.

One part of SAC's alert response took place at Little Rock AFB, Arkansas. The base was host to the 384th Bombardment Wing (Medium), a B-47 Wing, and its three Air Refueling Squadrons, all of which were equipped with KC-97 tankers. During those critical days in October 1962, the base was on full alert. Its B-47s were dispersed to other bases and even to commercial airports. For example, Boston Logan Airport, Massachusetts, and Memphis Airport, Tennessee, received numbers of B-47 bombers that were on 24-hour alert. In fact, SAC dispersed 183 B-47 bombers to 30 airfields, military and commercial.

The aircraft were dispersed because SAC felt that if the Soviet Union attacked the U.S. with missiles, it would target the B-47 bases first. Thus, by dispersing the B-47 bombers, there was a much greater likelihood some would survive and be able to fly missions. During the crisis, KC-97 tankers were also on 24-hour alert. At Little Rock AFB, maintenance crews worked around the clock to make sure the tankers would all be available if called upon to refuel B-47 bombers. In any event, the B-47 bombers were not launched during this period, and so the KC-97 tankers were not needed.

The KC-97 as an Airlifter

The KC-97 was by no means the only airlifter in the USAF. There were also such aircraft as the Fairchild C-119 and the Douglas C-124.

But because of the size of the KC-97 fuselage and its internal space, it could be quickly converted to haul cargo or troops. Thus, at times, the tanker played an important role when military needs required the fast deployment of troops or materiel. While a number of C-97 cargo aircraft had been manufactured for use as either cargo or passenger transports, at times KC-97s were also employed in that role, depending on the needs of the Air Force.This additional capability as an airlifter made the KC-97 that much more valuable to the USAF.

The KC-97 Goes to the Air National Guard

As SAC began receiving the new KC-135 jet tankers, it also initiated the transfer of a portion of its total KC-97 fleet to Air National Guard Refueling Squadrons. A total of 81 KC-97L tankers were eventually transferred to eleven Air National Guard Refueling Squadrons, some of which were located in or near metropolitan areas.

The first such transfer took place in 1959 when six KC-97L tankers were moved to the 126th Air Refueling Wing, which was based at Chicago O'Hare International Airport.

The KC-97s operating in Air National Guard units were used to refuel TAC fighters. Among these were A-10 fighter-bombers. Due to a peculiarity of the A-10, only the KC-97 could refuel that particular aircraft. That's because the A-10's airspeed was too slow for the KC-135 to refuel it. If a KC-135 did refuel an A-10, the tanker had to deploy both flaps and landing gear.

Operation Creek Party

Operation Creek Party is believed to have been the first mission employing the Air Force concept of Total Force. This concept utilized Reserve and National Guard aircraft to assist active duty units on actual missions.

In 1967, the Air Force launched Operation Creek Party. This operation involved the refueling of tactical fighter aircraft flying in Europe. The main reason for this mission was the continued threat by the Soviet Union in Europe, particularly in Germany.

The mission involved 11 Air National Guard units assigned to the 134th Air Refueling Group. The units operated out of Rhein-Main AB, Frankfurt, Germany, as part of U.S. Air Force Europe (USAFE). The units remained on station at Rhein-Main for 14 days each and then rotated back to their home base in the U.S.

The KC-97 tankers assigned to the 126th Air Refueling Wing, based at Chicago O'Hare International Airport, were among the units dispatched to operate as part of USAFE, the Air Force Command for Europe, which was also based at Rhein-Main AB, Germany. The operation lasted for 10 years, until 1977. In addition to the actual refueling carried out in Europe, every time a KC-97 rotated to Germany, regardless of its home base in the U.S., it had to fly a trans-Atlantic round trip. This added significant flying hours to these Air National Guard tankers.

This was the first time such a coordinated operation was employed. During those ten years, the KC-97s flew 6,500 sorties, performed 47,000 hook-ups, and offloaded 137 million pounds of jet fuel. These statistics are a testament to how the KC-97 fleet carried out its responsibilities.

The end of Operation Creek Party coincided with the retirement of the KC-97 as a tanker in the Air Force.

A Strange Landing

One day in the early 1960s, I was part of a Civil Air Patrol practice mission that took place at Rockford Airport in Rockford, Illinois. A number of us were standing just inside a hangar trying to avoid the sun on an extremely hot and humid July day when we heard the deep bass sound of an approaching aircraft. It was a KC-97 based at Chicago O'Hare Airport, practicing takeoffs and landings at Rockford.

The KC-97 made its approach with gear and flaps extended and then was unable to touch down. We all guessed that this might have been due to a combination of a slightly too high approach speed and very hot air rising from the concrete runway. The tanker floated down the runway only a few feet above it. Halfway down the runway, takeoff power was applied, and the tanker climbed out and went around the pattern. The second time it managed to touch down and made a touch-and-go maneuver. We all thought it was a strange sight to see that very large airplane over the runway but only a few feet above it.

KC-97 Accidents

During their operational career, KC-97 tankers suffered a number of accidents—some fatal, some not. There is no question that the demands of bomber fueling could create potential dangers. Weather was always a factor, as was maintenance. The engines of the KC-97 generated a great deal of horsepower, but these engines were also of a very complex design. Thus, the maintenance of the engines was always somewhat of a challenge. Beyond that, there was the human factor of a crew misjudging a landing. There was also the possibility of human error on the part of the receiver aircraft getting too close to the tanker, especially at night. But considering the thousands of fueling hook-ups by the KC-97 fleet, the accident rate was actually amazingly low.

A list of all the accidents sustained by the KC-97s during their career is shown in Appendix A.

The Legacy of the KC-97

SAC was perfectly aware that the KC-97 was strictly an interim tanker until a jet tanker became available. But the combination of the KC-97's fuel-carrying capacity with the newly developed flying boom meant that, for the first time, aerial refueling would become routine. Many of the procedures developed during the lifetime of the KC-97 would later become the nucleus of the next generation of tankers, the KC-135s. It is also worth considering that the combination of the B-47 bomber with the KC-97 tanker enabled SAC to routinely fly missions to the edge of Soviet airspace. It is likely that during the Cold War in the 1950s, those long-distance missions were a constant reminder to the Soviet government of the extreme danger of any expansionist activity on its part.

When the KC-97 was replaced by the KC-135 as SAC's tanker, it moved to the Air National Guard and TAC and continued to carry out air refueling missions for ten more years.

The KC-97 did its job and did it well, and should be remembered as a major step forward in the history of aerial refueling. Beyond that, the reader should note that before aerial refueling became routine, no rapid global response was possible, either for combat missions with bombers and fighters or for humanitarian relief missions.

However, having KC-97 tankers refuel B-47 medium bombers and later B-52 heavy bombers was never an ideal solution because those bombers were always close to their stalling speed while being refueled. The answer was a jet-powered tanker.

It is safe to say that the KC-97 represented an important and perhaps vital step in the further development of aerial refueling. But the KC-97 was getting close to the ultimate limits of reciprocating engine development. Turboprop-powered tankers might have been a possible interim step before moving to all-jet tankers. However, turboprop aircraft of the time were not that much faster than reciprocating-powered aircraft. When commercial airlines began looking at jet-powered aircraft, it helped instill in the minds of military leaders, such as Curtis LeMay of SAC, the distinct possibility of a jet-powered tanker. That idea became a reality as described in the following chapter. Aerial refueling was about to reach the next level of maturity.

CHAPTER 10
BOEING AND THE KC-135

In April 1952, the Boeing Company Board of Directors and its CEO, William Allen, decided to take a $16 million gamble and began building a new jet transport and demonstrator. In 2025, that original 1952 cost of $16 million would translate to at least $200 million, if not more. As we know today, that gamble paid off handsomely!

Fig. 10-1 Boeing CEO Bill Allen

Fig. 10-2 The Boeing 367-80 Prototype

In the 1950s, when Boeing bet on that ambitious and potentially risky project, it probably had some reasons to expect the project to succeed. Boeing knew the USAF had indicated it was looking for a jet tanker to replace the KC-97 tanker. American Airlines had also shown interest in potentially becoming the first U.S. airline to acquire passenger jets.

However, at the time, it was not likely that anyone at Boeing foresaw that the project would eventually lead the company to become the world's largest provider of passenger jet airplanes. The first result of the ambitious project was a one-off airplane identified at Boeing as the Model 367-80. It quickly became known simply as the Dash 80. This single airplane, built as a test bed and demonstrator, was considerably smaller than either the first passenger version, the Boeing 707, or the USAF KC-135A tanker. The Dash 80 was smaller and weighed less than either of the other two versions, but was still a large airplane with a maximum gross takeoff weight of 190,000 pounds.

Fig. 10-3 Boeing 707-100

The Dash 80 was flown all over the world on demonstration flights and proved to be an excellent ambassador that in time paid off its initial investment many times over. There is an interesting event that occurred in the summer of 1955 and that took the breath away from hundreds of thousands of Seattle residents gathered on a holiday weekend on the shores of Lake Washington to watch a Gold Cup Regatta motorboat race of unlimited-class hydroplanes. Boeing's chief test pilot, Tex Johnston, had been scheduled to fly the Dash 80 over the lake to show it off. He certainly did that! Johnston came over the lake at 500 feet of altitude, and as the huge crowd below watched in disbelief, slow-rolled the big plane. He flew on by, turned around, and on the return flyby slow-rolled it again! In the crowd below was Bill Allen, President of Boeing. Allen must have come close to having a stroke when he saw Boeing's very expensive demonstrator slow-rolling as if it were a small fighter aircraft. When Johnston landed back at Boeing Field, there was an urgent message that Allen wanted to see him immediately! No one knows what was said, but it is likely that Johnston came very close to losing his job. However, he must have explained to Allen that he knew for a fact that if properly executed, a slow-roll would not in any way put excess stress on the wings. Evidently, the explanation must have satisfied Allen because Johnston kept his job!

While the commercial airlines were initially hesitant to commit to a jet transport for their passengers, the USAF and SAC were extremely interested. Curtis LeMay already had a jet bomber in the SAC fleet, the B-47. He also had another jet bomber, the B-52, which began to be delivered to SAC in 1952. Both of these aircraft had difficulty refueling from the KC-97 tankers because their high stalling speed was close to the maximum speed of the KC-97. For SAC, the KC-135 represented the ideal solution to the problem.

In 1954, SAC had held a competition to see what aerospace company had the best design for a jet tanker. The race came down to the Boeing proposal for a tanker based on the Dash 80 prototype and a design from Lockheed known as the L-193 airliner. The fact that Boeing already had a demonstrator flying gave it a tremendous advantage. Furthermore, the Boeing entry could be delivered two years before the Lockheed aircraft. SAC declared the proposed Boeing tanker the winner, and the Air Force placed an initial order for 250 of the jet tanker. In time, the orders for the KC-135 would total 758 tankers. At Boeing, this new tanker project was identified as the Model 717.

When Boeing was first designing the initial 707-100 airframe, it saw this airplane as basically the same in both the Air Force and commercial versions. The SAC tanker was designed first. But when designing the commercial 707 version, Boeing ran into a major marketing problem. American Airlines, which was showing real interest in the 707, was adamant that Boeing widen the fuselage by four inches compared to the fuselage of the Model 717 tanker. This would permit six-across seating instead of five-across. Such a change would make the 707 considerably more profitable for American Airlines. During negotiations, the airline was quite insistent on that point. So Boeing felt it had no choice but to design two different fuselages: a narrower one for the tanker role and a wider one for the commercial version. The decision by Boeing was probably a very good one, since American Airlines became the initial customer of the 707, which then convinced many other airlines to also order the same airplane.

However, fabricating two models with a different fuselage diameter naturally complicated the manufacturing process to some extent. It meant that two fuselage fabrication lines would be required. The table below shows the different cabin widths of the KC-135 and the 707.

The KC-135 is a truly beautiful airplane. Sitting on the ramp, it looked as if it wanted to fly right then and there. The tanker was not the same airplane as the Boeing 707, even though they had an almost identical appearance. Both aircraft had the same 35-degree sweep-back wings and four jet engines in pods attached below the wings. The tanker fuselage had no side windows but did have a large cargo hatch on the left side. The major differences between the KC-135 and the two succeeding commercial versions are shown in the table below.

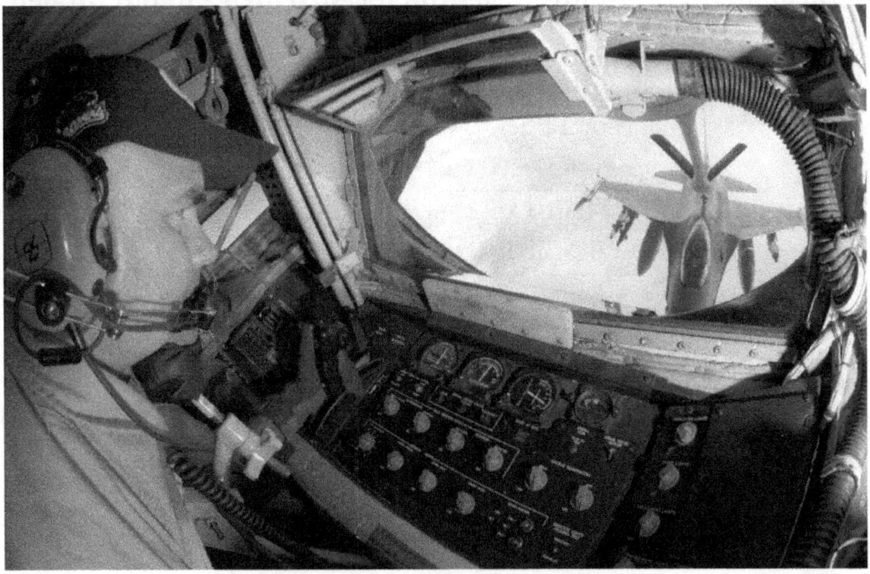

Fig. 10-9. KC-135 – Boom operator position

SPECIFICATIONS

	KC-135	KC-707-200B	KC-707-320B
Cabin Width	144"	148"	148"
Length	136' 3"	145' 1"	152' 11"
Tail Height	38'	41' 8"	42' 5"
Max to Weight	322,500 lbs.	257,340 lbs.	316,000 lbs.
Wingspan	130' 10"	130' 10"	145' 9"
35 Degree Sweepback on All Three Aircraft			
Engine Thrust	(X4)3,750 lbs.	18,000 lbs.	19,000 lbs.
Passenger Capacity	250	189	189
Cargo Capacity	83,000 lbs.	N/A	N/A
Performance Ceiling	41,000'	41,000'	41,000'
Range	8,000 miles	3,600 miles	5,200 miles
Cruising Speed: Approx. 530 Mph for All Three Aircraft Models			
Fuel Capacity	200,000 lbs. of transferable Fuel		
	Total Fuel Capacity 31,200 gal.		
	Main Tanks 62,738 lbs.		
	Auxiliary Tanks 126,419 lbs.		

A detailed description of the arrangement of the fuel tanks will help the reader better understand the fuel system of the aircraft.

1. Left wing outboard fuel tanks – 2,113 gal./14,167 lbs.
2. Right wing outboard fuel tank – 2,113 gal./14,167 lbs.
3. Left wing inboard fuel tank – 2,279 gal./15,270 lbs.
4. Right wing inboard fuel tank – 2,279 gal./15,270 lbs.
5. Left wing reserve fuel tank – 434 gal./2,908 lbs.

6. Right wing reserve fuel tank – 434 gal./2,908 lbs.
7. Forward body fuel tank – 5,800 gal./38,680 lbs.
8. Aft belly fuel tank – 6,378 gal./42,733 lbs.
9. Center wing fuel tank – 7,271 gal./48,718 lbs.
10. Upper lobe fuel tank – 2,174 gal./14,586 lbs.
11. Forward belly fuel tank – 5,800 gal./38,680 lbs.

Since the fuel could be transferred from any tank either for refueling or for use by the tanker, this gave the crew a great deal of flexibility. In addition, fuel could be transferred between tanks so as to maintain the proper center of gravity of the aircraft.

The KC-135 carried a refueling boom similar to that originally installed in the KC-97. In addition, it also had two wing-tip-mounted drogues plus a third drogue that trailed behind the boom. These three drogues are used when refueling fighters.

The KC-135 had a narrower cabin, a shorter fuselage, and was a considerably heavier airplane than the B-707-120B, as shown in the table above. However, it was not quite as big as the B-707-320B intercontinental version that came along shortly after the first B-707 and was designed to have the range necessary to fly across the North Atlantic Ocean non-stop.

As originally built, the KC-135 was powered by four Pratt & Whitney J57-P59W turbojet engines, each generating 10,000 pounds of thrust at take-off. These engines also had a feature designed to produce additional thrust during take-offs. This consisted of injecting 670 gallons of demineralized water over a period of two minutes. With water injection, the thrust per engine increased by 2,250 pounds to 12,250.

The one great advantage that the KC-135 had over previous tankers was its much higher cruising speed and longer range, plus its ability to carry a greater quantity of transferable fuel than the KC-97. The higher cruising speed made it far easier to transfer fuel to either jet bombers or jet fighters. The KC-135 also had a higher ceiling than the KC-97, which made it more likely that refueling could be carried out above clouds and storms. In short, the Air Force, in ordering the KC-135, got a tanker that represented a whole new dimension. The KC-135 was a much bigger airplane than the KC-97, and in addition to the advantages mentioned above, had far more space not just for fuel but also for troops, cargo, or medevac litters.

A comparison of the KC-97 versus the KC-135 is illuminating.

	KC-97L	**KC-135A**
Max to Weight	175,000 lbs.	322,500 lbs.
Transferable Fuel	14,980 lbs.	26,650 lbs.
Range	3,250	8,000NM
Ceiling	35,000’	41,000’
Cruise Speed	230 MPH	530MPH

For Boeing, manufacturing these three models simultaneously presented some real challenges because of their various differences. Nevertheless, Boeing not only managed to produce these three aircraft at the same time but was also building the B-47 and B-52 during the same period. This resulted in a tremendous growth of Boeing's manufacturing facilities and manpower.

The main cabin of the KC-135 could, if needed, carry 250 fully-equipped troops or 37 passengers, far more than the KC-97. The KC-135 could also be configured to carry either 31 medical litters or ambulatory patients. This additional capability simply made the aircraft even more valuable to the USAF because of its mission flexibility.

Operationally, the KC-135 was certainly a vast improvement over the KC-97. Aside from its carrying capacity and its cruise speed, the KC-135 had a much greater initial climb rate and was certainly more maneuverable. With this tanker, SAC finally had an operational aircraft that could easily refuel the B-47 and B-52 and eventually the B-58.

The KC-135 was built in different models as shown below.

KC-135A - 724

KC-135B - 17

C-135F - 12 (French Air Force)

Aircraft such as the KC-135 are built in batches. These might be a batch of 10 or 20 or more aircraft that are manufactured consecutively. Since these were military aircraft, the actual size of a batch depended on the amount of funding that the USAF and thus Boeing received from Congress in a particular fiscal year.

The first KC-135A Stratotanker came off the Boeing assembly line in Renton, Washington, on July 18, 1956. On the same day, Boeing also rolled out the last of the KC-97 tankers. The last KC-135 was delivered in January 1966, completing a continuous ten-year production cycle for that model.

The first KC-135A was delivered to Castle AFB, California. Initial operations concentrated on training pilots that had been flying the much slower KC-97s and were now flying at 550 mph on aircraft that were twice the size.

Boeing built a total of 732 KC-135A tankers, 17 KC-135B tankers, 45 C-135A or B transport aircraft that had no refueling gear; 14 C-135 aircraft for special military use; and an additional 12 tankers sold to the French Air Force. A grand total of 820 C/KC-135 aircraft were produced by Boeing. As of August 2025, 396 KC-135 tankers will remain in the Air Force inventory. In fact, as of that date, they will still provide the backbone of the USAF refueling fleet and are expected to remain operational until 2040.

Crews also began training for in-flight refueling of B-52 long-range heavy bombers which were also coming off the Boeing assembly lines at the same time. The early models of the KC-135A cost about $40 million each. As SAC converted each Refueling Squadron to the new KC-135A, the squadron's KC-97s were either transferred to Air National Guard units or were flown to the Air Force aircraft graveyard at Davis-Monthan AFB near Tucson, Arizona.

Initial KC-135 Problems

The flight tests of the new KC-135 revealed a number of problems as identified below.

Rudder Flutter

The KC-135A was not entirely free of problems. Early in the testing phase of the airplane, the flight crew wanted to test how much pressure was required to move the rudder. When pressure was applied to the rudder at high speed, the airplane began to flutter and oscillate. These continued until speed was reduced. Upon landing, it was discovered that part of the metal skin covering the vertical fin had been ripped off. Boeing attempted to correct the problem by installing a hydraulic damper at the top of the vertical fin. When this did not correct the problem, Boeing recommended that airspeed be limited to Mach 0.87 or 350 knots indicated airspeed (IAS). To further protect the vertical fin, Boeing applied additional skin over the upper part of the fin.

Dutch Roll

The KC-135 had a stability problem which came to be known as the "Dutch roll" and was common to all large aircraft with wing sweepback. The "Dutch roll" was clearly visible in flight, as the airplane could be seen to wallow slightly from side to side. This author experienced the phenomenon in the summer of 1959 while a passenger on a Pan American Airways flight from New York to Paris, France, aboard a Boeing 707-320 Intercontinental. Although the problem had been partially eliminated by that time, it was still visible if one knew what to look for. The problem was eventually solved by upgrading the autopilot software.

Boom Control

The refueling boom required high forces on the maneuvering stick in order to control it properly. Boeing's solution was to install hydraulic controls for the booms.

Lack of Engine Power

During the test phase, it was determined that engine thrust was insufficient for takeoffs at maximum gross weights, particularly at high-altitude airports or in very hot temperatures. This lack of thrust resulted in very long takeoff rolls. It was also shown that at those takeoff weights, if power was lost on one engine, the aircraft could become uncontrollable. The immediate solution was to install a water injection system. There were two such systems: one for the two starboard engines and one for the two port engines. Demineralized water was used, and the water injection could last for a maximum of two minutes. This added 2,250 lbs of thrust per engine. However, if one of the two water injection systems failed on takeoff, the aircraft would become unstable and almost certainly crash. In fact, a number of KC-135 crashes were caused by precisely that very thing. As a result, Boeing eventually redesigned the water injection system into an inboard-engine system and an outboard-engine system that was somewhat safer. By 1969, all KC-135A tankers had been retrofitted with the new system, which, however, was still not considered a good solution for the overall lack of engine power. Eventually, the KC-135A tankers were retrofitted with more powerful engines that provided greater thrust, as we shall see later in this book. It is interesting to note that, although the KC-135A tanker was considerably heavier than either the Boeing 707-200B or the 707-320B commercial airliners, it actually had less total engine thrust.

In those early days of jet engines, sufficient thrust was always a problem. This writer still remembers a clear example of this from August 1959. I was a passenger on a Pan Am Boeing jet. I do not recall if this was a 707-120 or a 707-320 Intercontinental. The flight was from New York City to Boston Logan and then non-stop to Paris, France. We landed at Boston Logan in the midst of a typical summer thunderstorm with a great deal of standing water on the runway. An hour later, we taxied out for takeoff. The cabin was completely full, and the aircraft must have been close to its maximum takeoff weight. Rain was still coming down in torrents.

The crew taxied to the takeoff position. I knew that the runway was close to 10,000 feet in length and it ended at the edge of Massachusetts Bay. The crew set the brakes, ran the engines up to full power, and then released the brakes and started the takeoff roll. I had been timing the takeoffs. When, at the 50-second point, we were still hard on the runway, I began to be very concerned. A few seconds later, we finally rotated and broke ground, just as the waters of the Bay appeared beneath us. I'll never know the truth, but it seemed to me that the combination of high ambient temperature, water on the runway, and marginal engine thrust all made that takeoff barely within safety margins. For the KC-135A, with less engine thrust, takeoffs must, at times, have been barely within safety margins.

The basic KC-135 was and still is an excellent aircraft design. But the original jet engines did not allow much margin for safety or error.

Fuselage Damage

The water injection system caused a totally new problem. When the water injection system was turned on during takeoffs, the noise level rose and became deafening, and it was discovered that the generated sonic waves were causing serious damage to the aft side panels of the fuselage. Boeing created a fix for the problem by applying additional skin to the areas that were being damaged.

Directional Control

Flight tests had already demonstrated a lack of sufficient directional control. To overcome this deficiency, Boeing designed a new vertical stabilizer 40 inches taller. This fix seemed to alleviate the problem considerably.

Yaw Control

It seems that no one realized it at the time, but aircraft with swept-back wings had a built-in problem if made to yaw violently. Such a maneuver resulted in a disastrous snap-roll. The KC-135, as well as the

707, was designed so that if too much stress was placed on the wings, the engine struts would fail and the engines would fall off before a wing failed. This result actually happened early in the 707 program during a training flight for Braniff Airlines lead captains, who were being trained by Boeing pilots. One captain, while following the training syllabus, pushed a rudder pedal with far too much enthusiasm. The 707 yawed and then snap-rolled and lost the two engines on one wing. Unable to adequately control the aircraft, the Boeing pilot in charge ordered the four Braniff pilots on board to the rear of the aircraft and the four Boeing employees to come forward. The aircraft crashed in a small river, and the Boeing personnel all perished. The Braniff pilots in the back all survived. After this crash, Boeing immediately revised the training syllabus, emphasizing the importance of minimizing yaw.

Why does a swept-wing aircraft react this way to a violent yaw? It is really very simple. When the aircraft is yawed, one wing produces maximum lift while the other wing's lift is reduced to almost zero. This unbalanced condition is what produces the snap-roll. While this is simple to understand, no one had ever foreseen the problem.

Stall Speed

The KC-135 had a relatively high stall speed. To increase the margin of safety, beginning in 1959, leading-edge flaps were added to all KC-135 tankers. These flaps deployed and retracted in conjunction with the standard trailing-edge flaps. The result was that the stall speed was lowered by six knots.

Conclusion

While it is undeniably true that the brand-new KC-135 had a number of serious problems, it is not uncommon for an aircraft of new design and with new technology to have such problems. The KC-135 represented a totally new generation of airplanes, and as such, problems had to be expected. However, the KC-135 was such an improvement over previous tankers that the work required to solve those problems was well worth the time and expense. The Air Force finally had a tanker that could far more easily refuel its jet bombers, fighters, and transports! As more and more KC-135 tankers were delivered to Air Force refueling squadrons, the USAF's global refueling capability increased. As described in the next chapter, the presence of the KC-135 made a very noticeable difference in air refueling operations.

Just as the KC-97 with the flying boom was a game changer, making refueling operations routine, so too was the KC-135. Equipped with the KC-135, SAC refueling squadrons now could reach their assigned stations much more quickly, transfer greater quantities of fuel, and bring the Jet Age to the refueling community.

CHAPTER 11

KC-135A OPERATIONS 1958-64

The initial SAC KC-135 operations were a continuation of the previous KC-97 flights, with the new tankers refueling SAC B-47s and B-52s. The main difference was that now, with the speeds of the tankers and the receivers being similar, refueling became considerably easier.

The nominal crew size on a KC-135 was five: aircraft commander, co-pilot, flight engineer, navigator, and boom operator. During this period, when the KC-135 was under the control of SAC, crews would fly a mission once per week. In addition, SAC also had some aircraft and crews on what they called "alert status." Crews on "alert status" were on alert one week out of every three. While on "alert status," the crew lived together in an alert shack located close to their aircraft and near a runway. The main mission of SAC, as most people are aware, was deterrence aimed at the Soviet Union. For this mission, B-52 heavy bombers flew continuously up to certain boundaries close to the Soviet Union. They might loiter there for a short time, then return to their home bases. The KC-135 tankers had the mission of refueling these bombers as required.

In the early 1960s, this writer spent a few days at Wright-Patterson AFB near Dayton, Ohio. At the time, the base had SAC units based there consisting of B-52 bombers and KC-135 tankers. Every night, around 2 A.M., I would be awakened by the sound of two B-52s taxiing to the active runway. One B-52 was the mission aircraft, and the other was a backup in case the first aircraft had to abort. The first B-52 would take off, and the second would taxi back to its parking space. Shortly thereafter, a KC-135 would depart heading for a refueling rendezvous with the B-52. It made for a noisy middle-of-the-night interlude.

The Cuban Missile Crisis

In October 1962, photos taken by U.S. reconnaissance flights clearly showed that the Soviet Union was building missile-launching sites on the nearby island of Cuba. These sites were designed to launch intermediate-range missiles that could reach the eastern third of the U.S. Then-President John F. Kennedy determined that this move by the Soviet Union was extremely dangerous and placed the U.S. armed forces on alert. One part of the military response, known as Chrome Dome, involved B-52 heavy bombers carrying nuclear missiles flying to the borders of the Soviet Union. During the crisis, 75 B-52 bombers were launched every 24 hours. Each B-52 required one or more KC-135 tankers to refuel it in order to complete its mission. These B-52 missions were undoubtedly designed so that each B-52 would register on Soviet radar, thus sending an unmistakable warning to the Soviet government.

The crisis lasted for 13 days and ended when the Soviet Union agreed to pull all offensive missiles out of Cuba.

The Chrome Dome Mission - 1960-68

Operation Chrome Dome was a direct result of the Cold War and was launched in 1960. As relations between the U.S. and the Soviet Union continued to deteriorate, Gen. Thomas S. Power, the commander of SAC, determined that it was essential that the Soviet Union be given a warning signal regarding its constant expansion. Operation Chrome Dome was that signal. It involved B-52D heavy bombers, loaded with nuclear weapons, flying close to the boundaries of the Soviet Union and orbiting there for a period of time. Gen. Power knew that placing the B-52s close to the Soviet Union's borders would register on Soviet radars.

In addition, the fact that these bombers were always carrying nuclear weapons was deliberately leaked to the Soviet Government.

The bombers were launched from various bases including Sheppard AFB, Texas; Homestead AFB, Florida; Barksdale AFB, Louisiana; Thule AFB, Greenland; and bases in the UK. The operation originally employed 14 bombers.

One of the tracks the B-52s followed began at Sheppard AFB, Texas. The bombers would fly northeast to New England and then on to Newfoundland, Canada. Then they would turn left and fly west of Greenland across the extreme northern area of Canada to northern Alaska, then turn left again and fly south along the Canadian and U.S. coast to near San Francisco and then fly southeast back to Sheppard AFB.

A variation of this route had the bombers fly to the east coast of Greenland before turning north and circumventing Canada. A third route ran across the North Atlantic to the Strait of Gibraltar, then across to Italy, and then around the Adriatic Sea before returning via the same route.

Other tracks approached the Soviet Union from different directions. In all cases, the B-52s would reach a certain point beyond which they were forbidden to fly unless they received an encrypted message that meant they should attack the Soviet Union and thus attempt to reach predesignated targets. These missions were often 24 hours long.

These missions each required a number of aerial refuelings by KC-135 tankers that would meet the bombers at prearranged locations. Without the tankers, Operation Chrome Dome could never have been carried out.

It is interesting to note that two movies were made and released that specifically referred to these flights. One such film was titled *Fail Safe*, released in 1964 and starring Henry Fonda, Dan O'Herlihy, and Walter Matthau. The other film, also released in 1964, was *Dr. Strangelove*, starring Peter Sellers, George C. Scott, and Sterling Hayden.

Operation Chrome Dome was not without its dangers. During its duration, four B-52 bombers crashed, killing all aboard. Because they were carrying nuclear weapons, the U.S. military rushed to those crash sites to recover the surviving bombs or their remnants and to make the sites safe from any radiation that might have been released.

Operation Chrome Dome ended in January 1968.

The 1961 Berlin Crisis – 1960-61

During the period 1960-61, the Soviet Union placed increased pressure on the city of Berlin, Germany, in an attempt to seize West Berlin from NATO countries. The U.S. Government, under the leadership of President John F. Kennedy, resisted this move with a number of military activities. In 1961, the crisis reached a peak. At that point, SAC went on alert, and B-52 crews were on 15-minute, round-the-clock alert. The KC-135 crews were on a similar alert status so that they could refuel the bombers on their way to the borders of the Soviet Union.

There are some statistics that show more clearly than words the level of activity by SAC during the month of October 1962 as part of Operation Chrome Dome. The B-52s flew 2,088 sorties totaling 47,168 hours of flying time. Each sortie required a minimum of two air refuelings and sometimes three. A fleet of approximately 133 KC-135 tankers was operating each day, off-loading 414 million pounds of fuel during the 31 days of the crisis. This sustained effort by the tanker fleet clearly demonstrated the value of the tanker fleet.

Throughout the 1960s, the B-52 fleet maintained steady pressure on the Soviet Union. During that period, the KC-135 fleet worked quietly behind the scenes to make the B-52 mission possible. In the meantime, a different U.S. military service became interested in A/R and decided to get involved.

CHAPTER 12
THE U.S. NAVY AND THE KC-130 HERCULES TANKER

In the late 1950s and early 1960s, the U.S. Navy and its member division, the Marines, became very interested in acquiring a tanker to refuel their aircraft. The choice made was for the Lockheed C-130 Hercules tactical transport. The C-130 family of airplanes in all its models represents a unique story in the annals of aviation. The very first Hercules, built for the U.S. Air Force, was introduced in 1956. Since that date, Lockheed has built various models of the C-130 continuously for an amazing 66 years. A number of foreign countries have also acquired C-130 aircraft over the years. Over 2,500 C-130s have been built to date, and a few are built each year, thus keeping the assembly line open.

There is an amusing story that supposedly took place while the very first C-130 was just beginning to be assembled. At a Lockheed Board of Directors meeting, an aide brought in a large model of the new aircraft and set it down on the conference table. Dead silence greeted the arrival of the model. Finally, one director said, "That's a pretty paint job on the model!" In other words, the directors were not impressed by the appearance of the airplane. It is true that the C-130 is not a pretty airplane since the design is basically very functional. But over the years and in all of its variants, it has proven to be a tremendous aircraft and has completely justified the design work of the Lockheed engineers.

The original version of the C-130 was a tactical transport/cargo aircraft powered by four turboprop engines. It had a straight wing mounted on top of the fuselage to allow ground clearance for the propellers. The fuselage sat close to the ground and had a rear ramp for cargo loading. The vertical rudder was quite high for excellent directional stability at low speeds. The nose featured a number of glass windshields to give the flight crew clear views in all directions. The basic Hercules design proved so useful that it was manufactured in many different variants and eventually exported to 17 foreign nations.

The U.S. Navy ordered the tanker versions of the KC-130 in various quantities as shown below.

Model Number	Ordered	Year Introduced-Retired
KC-130B	6	N/A-N/A
KC-130F	46	1962-2006
KC-130R	14	1976-2007
KC-130T	28	1983-2021
KC-130J	79	2004-Present
Total Ordered	173	

The KC-130F was developed from the B model and might be viewed as the baseline version of the KC-130 tanker. The KC-130T was a refueling tanker built for the U.S. Navy Reserves. It was similar to the KC-130R except that it had upgraded avionics. The KC-130R was used specifically by the U.S. Marine Corps. KC-130R tankers can transfer 8,000 gallons of fuel and have a mission radius of 1,000 miles. In June 1993, these aircraft were upgraded so that they had greater range.

Fig. 12-1 Side View of the KC-130F Tanker

Fig. 12-4 KC-130 Refueling Two Fighters

The extremely long production cycle of the Hercules resulted in succeeding models having various large and small upgrades. As shown above for the tanker models acquired by the U.S. Navy, there were no less than six different models, each with some variations. For that reason, the KC-130J model was chosen as representative of the Hercules tanker fleet to show its specifications. The basic dimensions of the airframe have remained more or less the same for all of the KC-130 models. The J model was also known as the Super Hercules and was first introduced in April 2004.

SPECIFICATIONS

- Wingspan – 132 ft. 7 in.
- Length – 97 ft. 9 in.
- Height – 38 ft. 10 in.
- Cabin width – 9 ft. 11 in.
- Max. T/O Weight – 175,000 lbs.
- Useful Load – 72,000 lbs.
- Fuel offload – 57,000 lbs.
- Power plants – 4 x Rolls-Royce AE 2100D3 turboprop engines, with 4,910 shaft horsepower each and with a Dowty 6-blade composite propeller.

CARGO CAPACITY

- 92 – passengers, or
- 64 – airborne troops, or
- 6 – cargo pallets, or
- 64 – litter patients & 2 medical personnel, or
- 2-3 Humvee vehicles, or
- 1 M113 armored personnel carrier

PERFORMANCE

- Cruise speed – 400 mph

- Range – 2,835 miles
- Ceiling – 28,000 feet with 42,000 lbs. payload

A minimum crew of 5 consists of 2 pilots, 1 crew chief, 1 loadmaster, and 1 boom operator.

The reader needs to understand that, because the KC-130 has been built over many years and in many versions, differences exist between the models. One change is that, over its manufacturing life, the engines of nearly all C-130s increased in horsepower.

All USN KC-130 tankers are equipped with two hose-and-drogue refueling systems, each of which is mounted at a wingtip. The hose-and-drogue systems are capable of refueling fixed-wing, rotary-wing, and tilt-rotor aircraft.

The J model has a 60,000-pound fuel capacity. This fuel can be allocated to either offloading or its own requirements. This provides a great deal of flexibility, depending on the particular mission. The fuel is carried in wing tanks plus two external fuel tanks located under the wings. When needed, an additional internal aluminum fuel tank with a capacity of 24,392 pounds of fuel can be installed in the cargo bay.

One of the great operational advantages of the Hercules aircraft is its ability to land and take off from very short runways. On occasion, when the Navy or Marines urgently required the delivery of cargo or the removal of injured personnel and no other aircraft were available, the KC-130s could perform this additional duty.

Conclusion

The KC-130 was a tough and dependable aircraft, and as it began serving the U.S. Navy, it quickly proved its value. The KC-130 turned out to be an excellent choice for the Navy and Marines, as we shall see in the next chapter.

CHAPTER 13
KC-130 OPERATIONS

The Marine Corps took delivery of its first KC-130F in March 1960.

The KC-130 tankers supported operations in a multitude of wars, including the Vietnam War, Operation Desert Shield, Operation Desert Storm, Operation Enduring Freedom, Operation Iraqi Freedom, and other Marine operations.

In Vietnam, one of the missions carried out by the KC-130 tankers was in support of the drawn-out siege of Khe Sanh. These tankers also saw duty in the various theaters listed above. The units that deployed to Afghanistan during the lengthy Operation Enduring Freedom had many refueling missions in often hazardous conditions.

Some of the KC-130 operations were particularly dangerous because they involved operating at low altitudes or in a live-fire war zone. In February 1968, a KC-130 was carrying cargo and fuel drums to the Khe Sanh area of Vietnam, where U.S. troops were under siege. While on approach to the Khe Sanh airfield, the tanker was hit by ground fire that caused the Number 3 engine to explode. The crew continued the approach and landed safely. But after landing, the aircraft exploded and came to rest on the runway in flames.

The tanker was carrying a crew of six and five passengers. Three of the crew and all five passengers perished, for a total of eight fatalities.

The Emergency of Raider 50

A vivid example of the ever-present danger during aerial refueling took place in 2020 not far from Yuma, California.

On September 29, 2020, a Marine Corps KC-130J tanker, call sign Raider 50, was on a training mission to refuel an F-35B fighter over California. The KC-130 pilots were Major Cory T. Jones and Captain Michael Wolff. For some unknown reason, the fighter collided with the tanker. The result was catastrophic. The pilot of the fighter ejected and landed safely while the fighter itself crashed. The damage to the tanker was also extremely serious. The collision ripped a hole in the tanker and resulted in the rapid decompression of the aircraft. On the flight deck, multiple alarms went off and caution lights lit up, which, added to the loud noise of the decompression, made everything more difficult. The collision destroyed both starboard turboprop engines but luckily did not damage the ailerons, elevators, or rudder. This permitted the crew to maintain control of the aircraft.

The crew contacted the FAA Los Angeles Air Route Traffic Control Center and declared an emergency. When any aircraft declares an emergency, air traffic controllers are quick to respond and try to render any assistance possible, including keeping other air traffic away from the aircraft that declared the emergency.

Raider 50 advised the Center of the collision, of having two engines out, and worse, of leaking fuel. That really caught the attention of the air traffic controllers! The fuel leaking from the damaged starboard wing caught fire, and Raider 50 trailed a long plume of fire and black smoke. The fire extinguishers for the right wing were activated, and the crew began to descend the aircraft to a lower altitude. With the aircraft so badly damaged, the crew was obviously anxious to get the KC-130J on the ground as soon as possible.

The crew looked around for a suitable place to land. Because the collision took place at a considerable altitude, the crew had a bit of leeway in which to find an airport. Major Jones decided they should be able to land at a small airfield outside of Thermal, California, called

Jacqueline Cochran Field in honor of the woman pilot who had led the women's group that ferried military aircraft in WWII. Raider 50 advised the Los Angeles Center of their intentions and proceeded to land at that small field. The plane crash-landed, but the entire crew was able to evacuate the aircraft. For their skill in dealing with the emergency, Major Jones and Captain Wolff were both awarded the Distinguished Flying Cross.

Tanker crews were always on the lookout for potentially dangerous situations such as that encountered by Raider 50.

Conclusion

The initial operations conducted by the Navy KC-130 tankers showed how capable the aircraft really were. In the years to come, the KC-130 would continue to prove its value. In the meantime, the Vietnam War had broken out. As this war escalated, the Air Force KC-135 tanker fleet became an integral part of air operations there, as described in the next chapter.

CHAPTER 14

KC-135 OPERATIONS – THE VIETNAM WAR

In 1964, the U.S., under the urging of then President Lyndon B. Johnson, began an undeclared war against North Vietnam and in support of South Vietnam. This war was fought under various self-imposed restrictions. Much of the war involved air operations. Initially, SAC had in mind basing B-52 heavy bombers in South Vietnam, the distance from Saigon, South Vietnam, to Hanoi, North Vietnam, being a moderate 715 miles.

To better understand the challenges faced by SAC during the Vietnam War, consider the following. The distance from Travis AFB in California to Andersen AFB, Guam, is 7,100 nautical miles, while the distance from Guam to Saigon, South Vietnam, is an additional 2,250 nautical miles. These kinds of distances meant that the SAC tankers were an absolute necessity. TAC fighters deployed to Southeast Asia required a number of air refuelings. Even B-52 bombers deployed initially to Guam required a precautionary air refueling.

This plan changed abruptly in November 1964. Five USAF B-57 Canberra medium jet bombers sitting on an airfield tarmac in South Vietnam were destroyed by a North Vietnamese air attack. SAC decided it could not risk basing its B-52 bombers in South Vietnam. Instead, they were based in Guam, and the tankers then became necessary for the B-52s to fly their missions due to the distances involved.

In early 1964, SAC deployed four KC-135 tankers to Clark Air Base in the Philippines. Called "Yankee Team Task Force," these tankers were to be on a temporary mission. On June 9, 1964, Yankee Team performed the first combat air refueling of the war, when it refueled eight F-100D fighters that flew a strike mission on Laos against the Pathet Lao in northern Laos. On June 22, Yankee Team was withdrawn from Clark AB, but then, after the Gulf of Tonkin incident on August 4, 1964, Yankee Team was re-established on August 5. This tanker force was then renamed "Foreign Legion" and supported tactical aircraft based at Clark Air Base. In time, the air refueling operations carried out during the Vietnam War became known informally as the "first tanker war."

As the Vietnam War continued to expand, more and more aircraft were deployed to Southeast Asia. As a result, in January 1965, a Tanker Wing was established at Kadena Air Base on the island of Okinawa, formerly Formosa. That wing would remain there for eight years. The tanker force in Southeast Asia was renamed "Young Tiger."

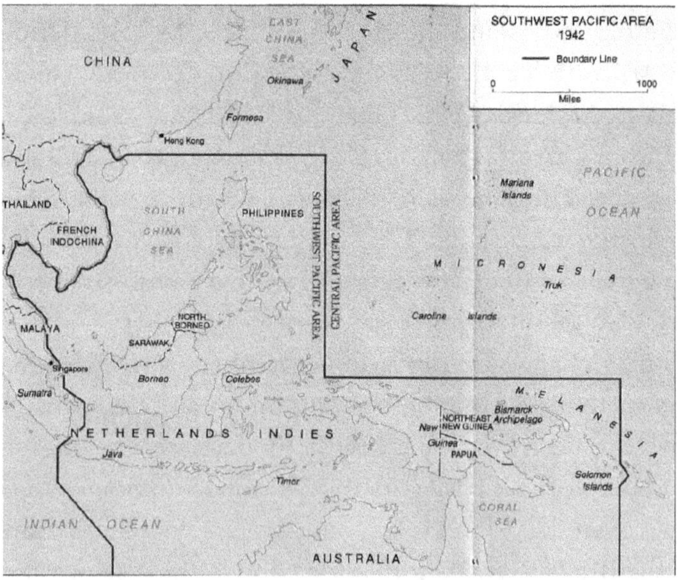

Fig. 14-2 Map Showing Location of Kadena Air Base, Okinawa, and Clark Air Base, Philippine Islands

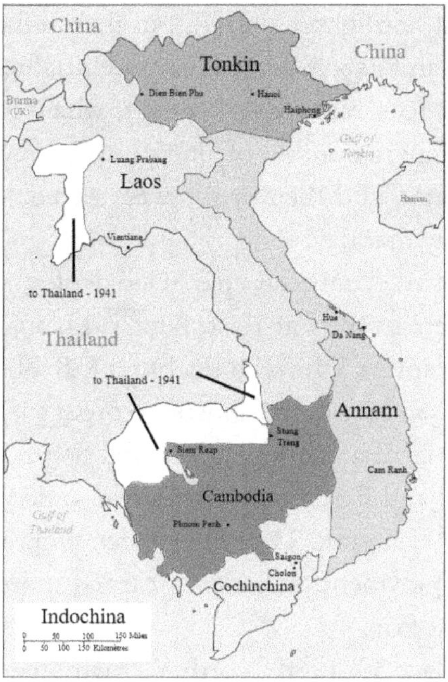

Fig. 14-3 Map of Indochina

On Feb. 12, 1965, 38 KC-135 tankers supported the deployment of 30 B-52 bombers to Andersen AB on Guam. The B-52s then began bombing missions against Viet Cong jungle locations. These missions were supported by the KC-135 tankers based at Kadena AFB. The missions against the Viet Cong began on June 18, 1965, and involved 30 KC-135 tankers refueling 27 B-52 bombers. The bombers were heavily laden with conventional bombs, and it was the tankers that made the 4,200-mile round-trip possible.

On Nov. 22, 1965, a USAF F-105 fighter, which had completed its mission, began losing fuel. A KC-135 from Kadena AB was dispatched to its aid, and the fighter managed to recover at Clark AB.

Another "save" occurred on Nov. 22, 1967, when a tanker from Kadena AB was on a routine mission over the Gulf of Tonkin refueling two F-104C fighter aircraft. This particular KC-135 was equipped with a boom-drogue adapter in order to refuel the F-104s. Some U.S. Navy aircraft arrived, also dangerously short of fuel. The tanker not only refueled the F-104 Starfighter aircraft, but then refueled two Navy F-4 Phantom fighters and two Navy F-8 Crusader fighters. In addition, the tanker passed fuel to a Navy KA-3 tanker, which in turn refueled another Navy F-8. Thus, a number of aircraft were prevented from having to ditch in the water, and their crews were saved. That KC-135 crew really earned its pay that day!

In early 1968, two events caused SAC to deploy more B-52 bombers and KC-135 tankers to the Far East. First, on January 23, 1968, North Korean warships captured the USS Pueblo, a U.S. Navy intelligence vessel. This event caused President Johnson to order a large force of aircraft to deploy to the Mariana Islands, Japan, and South Korea. B-52 heavy bombers, plus F-4 and F-105 fighters, were also deployed in a force totaling more than 200 aircraft. The tanker force supported and made possible the quick deployment, which was carried out under the name of Operation Combat Fox.

Then, on January 30, 1968, North Vietnam, together with the Viet Cong, launched a military operation known as the Tet Offensive. In response to this offensive, SAC deployed additional B-52 bombers and KC-135 tankers to the Far East under Operation Port Bow.

A third operation, Operation Commando Royal, had KC-135 tankers supporting the deployment of tactical aircraft to the Korean Demilitarized Zone. This latter operation resulted in an average of five tanker sorties per day.

The Tet Offensive caused the siege of Khe Sanh. Tankers and bombers began flying "Arc Light" missions as part of Operation Niagara. These missions were to relieve Khe Sanh, which they did. As a result, the average number of tanker sorties per day grew from 53 to 66. On April 1, 1968, President Johnson halted the bombing missions, thus allowing the tanker crews a respite. The sortie numbers are a clear indication of both the importance of the tanker sorties and the level of daily operations. Nevertheless, the tankers had to continue supporting other operations in South Vietnam and Cambodia.

Although the tankers were based at many different bases, such as Guam, Okinawa, and the Philippines, a main base was established at U-Tapao Royal Thai Air Base in Thailand. There a force of 46 tankers and 88 aircrews flew almost daily.

When flying in support of B-52 bombing strikes, 3 or 4 tankers would fly together with the B-52 leading the mission. The remaining tankers would remain a mile or so behind the bombers and slightly higher. They would then refuel the bombers both to and from their targets. For the refueling of fighters and reconnaissance aircraft, refueling areas were set up over the Gulf of Tonkin, Thailand, Laos, and South Vietnam.

In a separate effort, USAF KC-130P tankers refueled search and rescue and special operations HH-3 Jolly Green Giant helicopters, thus greatly extending their range.

In 1972 the air war became more intense. In 1972 and 1973, SAC sent more tankers to the Philippines and to Thailand. On August 15, 1973, a final air strike over Cambodia signaled the end of the air war in Southeast Asia. Throughout the Vietnam War, from 1964 to 1973, the tankers operated continuously for nine years and two months. During that lengthy period, the tanker fleet flew 194,687 sorties, 911,364 total flying hours, transferred approximately 1.4 billion gallons of fuel, and completed 813,878 refueling contacts. During this same period SAC lost four KC-135 tankers, two on takeoff and two during landings. No one will ever know how many aircraft and aircrews, Air Force, Navy, and Marine personnel, were saved due to the presence of the KC-135 tankers.

Conclusion

It is no exaggeration to say that without the tanker force the air war in Southeast Asia would probably have been impossible to carry out.

In its original role the KC-135 was designed to refuel SAC bombers and TAC fighter aircraft. But over the long span of its operational life, other operational roles were added. Some of these will be described in the following chapters.

CHAPTER 15
KC-135 OPERATIONS 1973-1981

During the period 1973 to 1981, the KC-135 fleet not only carried out its routine fueling operations but also engaged in an overseas operation as detailed below.

Operation Nickel Grass – October 12 – November 14, 1973.

Operation Nickel Grass began October 12, 1973, and ended November 14, 1973. It was organized in an extremely short time. This operation was the result of the combined surprise attack by Egypt and Syria on Israel on October 8, 1973. This day was Israel's most religious holiday, known as Yom Kippur or the Day of Atonement; thus, the ensuing war became known as the Yom Kippur War. Then-U.S. President Richard Nixon, upon being informed of the attack on Israel, immediately authorized the U.S. military to begin air shipments of war materiel to Israel.

Israel had already started sending commercial Boeing 707 aircraft of the national airline, El Al, to U.S. bases to begin picking up some of this materiel. But these aircraft were too few in number and were configured for passenger use. As a result, U.S. Air Force C-141 and C-5A military transports began an emergency airlift to Israel. This airlift was complicated by the fact that Portugal was the only European country that allowed U.S. military aircraft to overfly its airspace.

Fig. 15-1 Map of Israel and Surrounding Area

Accordingly, the former WWII base on the island of Lajes, in the Portuguese Azores, was reactivated to act as a refueling base for the emergency airlift. The distance from U.S. bases to Lajes in the Azores was approximately 3,900 miles. From Lajes to Israel, the distance was an additional 3,500 miles. These distances were far too long for the heavily loaded C-141 and C-5A airlifters coming from bases in Texas. These distances were even more prohibitive for short-range A-4 Skyhawk and F-4 Phantom fighters coming directly from the factory in St. Louis, Missouri. These fighters were also meant for Israel. Furthermore, after leaving the Azores, all these aircraft had to fly directly to the Strait of Gibraltar and then had to follow a very narrow corridor across the Mediterranean Sea to Israel.

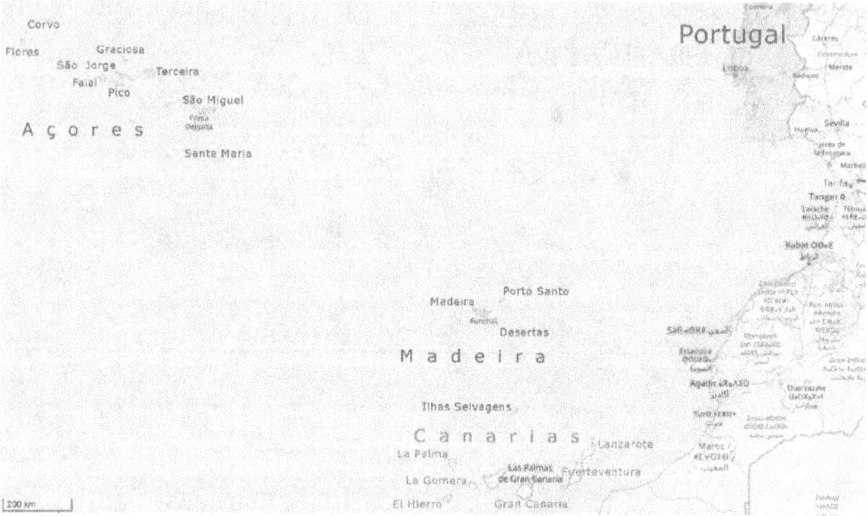

Fig. 15-2 Map of Azores and Portugal

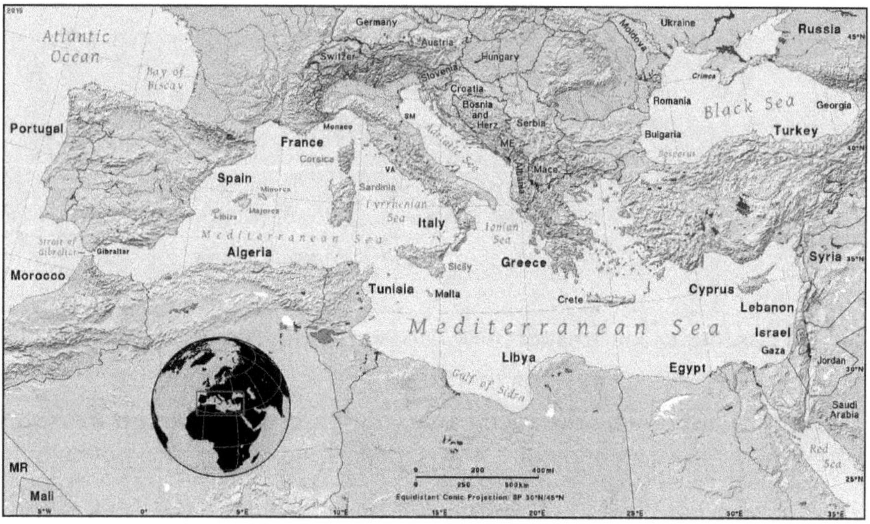

Fig. 15-3 Map of Mediterranean Sea

On October 12, KC-135 tankers of the Strategic Air Command left Pease AFB, New Hampshire, and flew to Lajes AB, Azores. The Azores are 800 miles west of Portugal. Thereafter, the KC-135 tankers were used to refuel the airlifters and fighters crossing the Atlantic Ocean on their way to Lajes, as well as those en route to the Strait of Gibraltar. From the Azores, the airlifters were probably refueled once more in the vicinity of Gibraltar. For the fighters, it was a different story. They were refueled by U.S. Navy tankers operating from three Navy carriers: one west of Gibraltar, one south of Sicily, and one near the island of Crete. The total airlift was a complete success, and not one aircraft was lost.

This successful transfer of fighters to Israel using tankers of both the Air Force and the Navy was the first such instance of a combined effort during an operation. It clearly showed, once again, the far-reaching effect of tankers when the U.S. felt it needed to project its military might far from its shores. It could be said that by now tankers were part and parcel of the U.S. military strike organization without which any military effort would have been either greatly hampered or have been impossible.

During this period, except for wartime operations, the KC-135 crews flew about once per week.

At the time of Operation Nickel Grass, it was rumored that the U.S. and the Soviet Union maintained secret communications. The reason given was fear of mid-air collisions over the Mediterranean Sea since Air Force aircraft supplying Israel were flying from west to east while military support by the Soviet Union to Egypt was flying north to south. In addition, commercial flights originating in Israel and flying west were under top cover by Israeli fighters for at least the first 100 miles west of Israel. Any mid-air collisions could easily have resulted in a confrontation between the U.S. and the Soviet Union.

Operation Nickel Grass showed that if U.S. Air Force airlifters, such as the C-5 and C-141 aircraft, could be refueled in the air, then the U.S. ability to project its force far from its shores would be greatly improved. As a result, these airlifters were eventually equipped with fueling receptacles located just behind the cockpit so that they could refuel from tankers using the flying boom system.

During this period, the KC-135 tankers also flew routine missions in support of fighter aircraft deployments. Col. Mark E. Smith, USAF, (Ret), remembers a very long ferry mission in which he flew an F-4 Phantom fighter from Lakenheath AB, Great Britain, to Nellis AFB, near Las

Vegas, Nevada. The flight took 13.5 hours. He was accompanied by a KC-135 the entire way. During the crossing of the North Atlantic, the F-4 made frequent refuelings so that, in case it had to divert to the closest airfield because of an emergency, it had enough fuel.

By this time it had become a routine part of planning for all Air Force operations to consider air refueling as a normal part of such planning. As the KC-135 fleet became older, the Air Force determined that a new tanker model was really needed. The following chapter introduces this new aircraft.

The Coronet Missions

Following World War II, the U.S. Air Force found that at times it was necessary to ferry fighter aircraft to distant bases in both Europe and the Far East. Initially, this requirement was directly associated first with the Cold War and then with the Korean War.

While fighters had been ferried to bases in such countries as England and Japan, these ferry missions were very lengthy and potentially dangerous. They required many stopovers, and the necessary crew rest periods meant that moving a fighter squadron on such missions could take weeks to complete.

With the introduction of KC-135 tankers, a fighter could now cross the Atlantic or Pacific Ocean in mere hours instead of days. This new ability prompted the USAF to create a program known as Coronet Missions. There are various types of Coronet Missions, but one involves the movement of fighters across either of these two oceans. Bombers, airlifters, and other such large aircraft are not part of these Coronet Missions since they have much greater range than fighters. This type of Coronet Mission began in 2020. It was so successful that Coronet Missions for fighters became a standard part of USAF operations when the movement of fighters was required.

For Coronet Missions, tankers operate somewhat differently than on other missions. On typical refueling missions, tankers wait in orbit, refuel the receiver(s), and then either depart the area or remain in orbit to refuel another receiver(s). On some Coronet Missions, tankers will wait in orbit for the receivers, refuel them while continuing to travel along the receivers' track, and then accompany the receivers on their flight. This enables the tanker to offload additional fuel to the receivers at various times, as well as to provide navigation and communication assistance. Furthermore, if the receiver(s) have to divert or ditch in the ocean, the tanker can call for a Search and Rescue (SAR) mission without delay.

Coronet Missions require as much as a month of preparation and planning to ensure that the required number of tankers will be on station as needed to meet fighters and transfer the required amount of fuel.

Coronet Mission tankers normally refuel fighters at altitudes of 24,000 to 28,000 feet. For A-10 attack aircraft, the refueling is done at 17,000 to 20,000 feet.

Fighters flying East across the Atlantic with destinations in Europe or beyond recover at RAF Mildenhall/Lakenheath, England; Spangdahlem AB, Germany; Morón, Spain; or Lajes Field, Azores. Included in these missions are fighters that are flying East over the Mediterranean Sea to the Middle East.

Flying West from the Continental United States (CONUS), fighters are refueled in the same way between CONUS and Alaska and then further West.

A critical part of the planning of these missions is weather, which, in extreme cases, can result in the postponement of a mission.

While KC-135 tankers handled these Coronet Missions for many years, when KC-46 tankers became operational they also flew these Coronet Missions.

Conclusion

This routine operation of air tankers enabled the USAF to dispatch fighters to distant areas much more quickly and safely.

CHAPTER 16
THE KC-10 EXTENDER

The introduction of a new and different tanker into the U.S. Air Force inventory in 1980 was the direct result of perceived shortcomings of the then-current KC-135 tanker fleet. In the late 1960s and early 1970s, during the Vietnam War, the KC-135s were operated at their maximum level. Then, in 1973, as a result of the Yom Kippur War in the Middle East, further deficiencies of the tanker fleet came to light.

Another reason why the Air Force wanted additional tankers was that by the late 1970s, the older KC-135s were beginning to require more and more maintenance to keep them in operation. This necessary maintenance was becoming increasingly expensive.

These events convinced the USAF of the need for a new and larger tanker that could also double as an airlifter for large military cargo items. Accordingly, a competition was initiated involving the Boeing 747, the McDonnell Douglas DC-10, and the Lockheed 1011 Tristar. The winner of this competition, announced in December 1977, proved to be the McDonnell Douglas DC-10.

In its original commercial version, the DC-10 had some major design flaws which resulted in a number of crashes and fatalities. By the time the USAF ordered the tanker version, it appears all those design problems had been corrected.

The aircraft that the Air Force purchased was an outgrowth of the DC-10F freighter version, which in turn was developed from the original DC-10. A comparison of these three models shows the growth from one to the next.

	DC-10	**DC-10F**	**KC-10A**
Wingspread	153’ 10”	165’ 4”	165’ 4”
Length	182’ 3”	181’ 7”	181’ 7”
Max to Weight	430,000 lbs.	555,000 lbs.	590,000 lbs.
Range	3,500 NM	5,200 NM	3,800-4,400 NM
Ceiling	42,000’	42,000’	42,000’
Cruise Speed	Mach 0.82	Mach 0.82	Mach 0.82
Engines	multicolumn{3}{l}{3x GE F103-50C2 turbofan jet engines, 52,000 lbs. thrust each}		

Although the KC-10 resembled the DC-10 to a large degree, it was a truly different aircraft. As shown above, the maximum gross takeoff weight was much greater. The main deck floor was reinforced to handle military cargo. Also, the KC-10 had extra fuel tanks and a more complex piping system that allowed great flexibility as to whether the fuel was used to power the tanker or offloaded to receivers.

The KC-10 is a wide-body aircraft powered by three jet engines: one under each wing and one in the tail just below the vertical fin and rudder. It had a far larger fuselage than the KC-135 and a maximum gross take-off weight almost twice that of the KC-135. Furthermore, its later-model jet engines enabled it to operate from shorter runways. All of this was made possible by the fact that the jet engines available in the 1980s were far more powerful than those available for the KC-135 in the late 1950s. In addition, the KC-10's greater engine power and larger fuselage, compared to the KC-135, enabled it to carry much more fuel and heavier cargo loads. The KC-10 can take off safely at maximum gross take-off weights, except in cases of extreme heat, where the take-off weight must be reduced. This capability contrasts with the KC-135, which had limitations on take-off weights due to insufficient engine thrust, also known as a power-to-weight ratio.

The original order for the new tanker was for 12 aircraft, but this was later upgraded to 60 aircraft. The first KC-10 was delivered to the Air Force in 1981 and the last in 1988. However, there are only 59 tankers in the fleet. That is because in November 1987, a KC-10 undergoing maintenance at Barksdale AFB, Louisiana, experienced an explosion and fire. The aircraft suffered major damage and was written off by the Air Force. One maintenance airman died as a result of the accident.

Fig. 16-1 Side View of KC-10

Because of its size, the KC-10 brought an entirely new dimension to aerial refueling. There is a significant difference between the KC-135 and the KC-10. The best way to demonstrate this is to compare a few basic metrics.

	KC-10	**KC-135**
Width	18' 2"	10' 9"
Max. Takeoff Weight	590,000 lbs.	322,000 lbs.
Fuel Capacity	365,000 lbs.	200,000 lbs.
Range	4,400 miles	1,500 miles
Max. Cargo Payload	170,000 lbs.	83,000 lbs.

Engine Thrust (per engine)	52,500 lbs. x 3	10,000 lbs. x 4
Total Engine Thrust	157,500 lbs.	40,000 lbs.
Cruise Speed	Mach 0.82–0.84	Mach 0.82–0.84

KC-10 Performance

- Service Ceiling: 42,000 feet
- Cruise Speed: Mach 0.825

Note the significant difference in total engine thrust between the KC-135 and the KC-10. As a result, the KC-10 could operate safely from shorter runways than the KC-135. Note also how much more fuel the KC-10 carried than the KC-135. Finally, when the KC-10 was employed as an airlifter, it had much greater internal space and could lift a far greater payload. Thus, the reader can see how the KC-10 was able to accomplish much more than the KC-135. The fact that the KC-10 was a wide-body aircraft meant that it had far more cargo space.

The basic design of the KC-10 is that of the DC-10-30CF, a cargo version of the DC-10. However, there are a number of differences from the commercial cargo version. Most of the upper lobe windows and lower lobe cargo doors were removed. An improved cargo handling system was installed, as were military avionics. The KC-10, having been designed and built more than twenty-five years later than the KC-135, profited from having an improved flight control system and better communication equipment installed.

The Refueling System

In order to perform its refueling mission, various equipment had to be added. The original DC-10-30CF aircraft had three fuel tanks in each wing. For the KC-10, three additional fuel tanks were installed in the lower lobe of the fuselage beneath the main floor. One is located under the forward cargo compartment, one in the center wing area, and one under the rear compartment.

Then, a McDonnell Douglas Advanced Aerial Refueling Boom was installed beneath the rear lower lobe. This boom differs from Boeing's flying boom in that it has a control surface system that replaces the V-tail of the Boeing flying boom. In addition, the KC-10 has a hose and drogue system on the starboard side of the rear fuselage to refuel fighters. The boom and the hose-and-drogue system constitute the centerline refueling system. Its operation is controlled by the boom operator, located in the rear of the lower lobe, who has a wide window through which to monitor the refueling operation. The two centerline systems cannot be used simultaneously. Refueling could be carried out with the boom first or the drogue first; it did not matter.

The last twenty KC-10s built also had wingtip-mounted pods that added two more hose-and-drogue locations for multiple refueling operations of fighter aircraft. Eventually, all KC-10 Extender tankers were outfitted with wingtip drogues. These could be installed or removed depending on the mission and required only two hours for the change. When the wingtip drogues were used, the centerline boom and/or drogue had to remain inoperative. The wing drogues are also controlled by the same boom operator. With this multiplicity of refueling systems, the KC-10 had the ability to refuel not only USAF bombers and fighters but also Navy, Marine Corps, and most aircraft of allied countries. The KC-10 could also transfer fuel to other Air Force tankers, including another KC-10. As a tanker, the KC-10 could carry up to 356,000 pounds of fuel. In its secondary role as an airlifter, the KC-10 could carry 75 personnel plus 246,000 pounds of cargo. The cargo was stored in the upper lobe, which sits quite high above the ground. Thus, cargo had to be lifted to that height by ground equipment.

The KC-10 fleet was originally assigned to SAC. As such, it operated from two main bases: Travis AFB on the West Coast and McGuire AFB on the East Coast. After the USAF completed its internal reorganization in 1992, the KC-10 fleet became part of the Air Mobility Command and remained at these two bases. The assignment of the KC-10s to these coastal bases directly reflected the aircraft's significantly longer range and larger payload capabilities. The KC-10's extended range led to the fleet's involvement in many long-range missions.

During the first three years that the KC-10 was in service, a full one-third of its flight hours were dedicated to humanitarian missions. The KC-10's range also enabled the tanker to handle missions in Europe and the Middle East that required overflying countries which had denied landing rights to U.S. military aircraft.

Training missions for the KC-10 all involved refueling. Actual missions included combined cargo and refueling operations. About 15% to 20% of missions were combined operations. However, missions involving cargo required at least 6 pallets of cargo to be on board.

Conclusion

The KC-10 proved its value over the years. Not only was it an excellent tanker, but its wide-body airframe was a great asset on many occasions. A new organization was poised to take over the tanker function of the Air Force, as we shall see in the next chapter.

CHAPTER 17
THE AIR MOBILITY COMMAND

In 1991, the Air Force underwent a major reorganization. As part of this significant change, a new command was created called the Air Mobility Command (AMC). AMC brought together two former organizations: the Military Airlift Command (MAC) and the tanker force, which had been part of the Strategic Air Command (SAC). AMC was established on June 1, 1992, and is one of the USAF's major commands. AMC Headquarters is located at Scott AFB, which is east of St. Louis, Missouri, in southern Illinois.

The airlift portion of AMC is made up of C-5 Galaxy, C-17 Globemaster III, and C-130 Hercules aircraft. The tanker fleet is made up of KC-135 Stratotankers and KC-10 Extender tankers, all of which could, if called upon, do double duty as airlifters.

The mission of AMC is to provide global air mobility in support of the U.S. armed forces. It has a secondary role to provide humanitarian aid both in the U.S. and around the world. The airmen who make up AMC are active duty, Air National Guard, and Air Force Reserve personnel. If, at any time, AMC needs additional capabilities, it can call upon the civilian flight crews and aircraft of the Civil Reserve Air Fleet, which are drawn from the major U.S. airlines.

AMC Organization

AMC's refueling aircraft are located at many bases throughout the U.S. The list below shows the location of these bases. The type of tanker at any base can change as KC-135 Stratotanker aircraft are retired while new KC-46A Pegasus tankers replace them. NOTE: The letters ANG shown for some bases indicate Air National Guard sites.

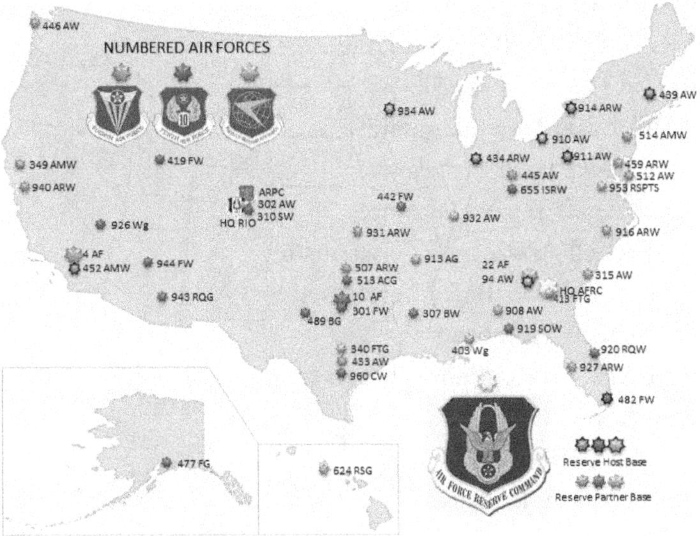

Fig. 17-1 Map Of U.S. Showing Tanker Base Locations

List of Tanker Bases
Travis AFB, California
McGuire AFB, New Jersey
Scott AFB, Illinois
March AFB, California
Grissom AFB, Indiana
Niagara Falls Air Refueling Squadron, New York
Seymour Johnson AFB, North Carolina
MacDill AFB, Florida
McConnell AFB, Kansas
Tinker AFB, Oklahoma
Keesler AFB, Mississippi

Bangor ANG Base, Maine
Bradley International Airport, Connecticut
Beale AFB, California
Sumpter Smith ANG Base, Alabama
Rickenbacker ANG Base, Ohio
Selfridge ANG Base, Michigan
Roland R. Wright ANG Base, Utah
Lincoln ANG Base, Nebraska
Pease ANG Base, New Hampshire
Goldwater ANG Base, Arizona
New Castle ANG Base, Delaware
General Mitchell ANG Base, Wisconsin
McGhee Tyson ANG Base, Tennessee
Fairchild AFB, Washington
Pittsburgh ANG Base, Pennsylvania
Sioux City ANG Base, Iowa
Key Field ANG Base, Mississippi
Forbes ANG Base, Kansas

A quick reading of this list of bases that have tankers clearly shows that AMC has aircraft in many different locations and that many of these refueling units are manned by Air National Guard personnel. The reader should also keep in mind that any tanker unit may be deployed overseas for a period of time.

In fact, on October 1, 2003, because of the Global War on Terrorism, AMC decided it needed to meet its new mission requirements with some reorganization. As a result, two (2) Expeditionary Mobility Task Forces were established. One is based at Travis AFB, California, and the other at McGuire AFB, New Jersey. In addition, as part of its peacekeeping and humanitarian missions, AMC tankers and airlifters have provided support in Afghanistan, Bosnia, Iraq, Cambodia, Somalia, Rwanda, and Haiti. In 2022, with the Russian invasion of Ukraine, AMC began a massive airlift to bring war materiel to that area in support of the Ukrainian Army.

Seventy years earlier, around 1950, the idea of supplying an ally with military supplies by air would have been laughed at. Now it is simply a part of the operations that air refueling has made both possible and routine.

Conclusion

The addition of the KC-10 Extender to the USAF tanker fleet brought three advantages. The KC-10s carried more fuel, had much greater range, and a far larger payload. Giving it the name "Extender" was definitely highly appropriate.

In the next chapter we shall see how the KC-135 fleet underwent changes as it got older and as it was relied on heavily.

CHAPTER 18

KC-135 CONVERSIONS & UPGRADES

Many of the KC-135 tankers were converted and also underwent a number of engine and other upgrades. These upgrades improved the KC-135's operational ability considerably. The conversions were made for different reasons. Because the conversions and upgrades ranged from hundreds of KC-135s to single aircraft conversions, and to avoid confusing the reader, only the major conversions and upgrades are shown here. The reader should also note that it was not uncommon for some of the KC-135 tankers to actually be converted more than once over their long operational life.

When a large batch of these tankers was either converted or upgraded, the model numbers changed. As previously noted, the USAF received a grand total of 732 tankers from the Boeing Company. The entire original KC-135 fleet was delivered between 1956 and 1964. All of these tankers were designated as KC-135A models. Below are listed some of the major KC-135 conversion and upgrade programs that took place over the years.

These various programs were all expensive and lengthy, but in retrospect, were extremely valuable as they increased the life of the KC-135 fleet by many years.

The KC-135 has been in service much longer than anyone originally anticipated. As a result, a number of conversions were authorized over the years to help the KC-135 maintain its operational usefulness. The original KC-135s were all equipped with the Pratt & Whitney J57 turbojet engine which produced 13,750 pounds of thrust each when using water injection on takeoff.

KC-135E

The various KC-135A problems detailed in Chapter 10 clearly revealed that while the airplane was potentially a great addition to the tanker fleet, it had a number of serious problems that needed to be remedied. SAC and the Air Force were anxious to find solutions for these various problems, with possibly the most serious one being that of insufficient engine power.

Curtis LeMay, in particular, wanted to see the KC-135 fleet upgraded to the point where it would meet the needs of SAC to a greater extent. As a result, in the late 1970s, there was a considerable amount of politicking in the Pentagon and Congress as the USAF attempted to find the funding necessary to get these engine upgrades. SAC indicated it might require a brand new tanker if it were to meet its future responsibilities. Boeing was enlisted to make recommendations as to what could be done to remedy the KC-135A's lack of engine power. In the end, the least expensive solution was to modify the KC-135A by simply installing more powerful engines.

A number of different options were suggested, but finally, the one accepted involved the Pratt & Whitney TF-33 turbofan engine. As it happened, a quantity of Pratt & Whitney TF33-PW-202 turbofan bypass engines that had been removed from retired Boeing 707 and 720 commercial airliners were available. In 1982, a major modification program resulted in 161 of the A-model tankers being retrofitted with these used engines. By that time, many JC-135 tankers had been transferred to Air Force Reserve and Air National Guard units. The TF33 turbofan engine had a thrust of 16,050 lbs, considerably more than the original JT3, which had a thrust of 10,000 lbs. The TF33 engine was 14% more efficient and allowed the tankers to offload 20% more fuel on long-duration missions. This group of re-engined tankers was designated KC-135E. In addition, they were also fitted with thrust reversers of the "clam shell" type. These engine reversers resulted in safer aborted takeoffs as well as shorter landing rollouts. As part of this modification, engine struts from retired Boeing 707 and 720 commercial transports were used to mount the retrofitted engines onto the KC-135s. With the more powerful engines, the water injection system was no longer deemed necessary and was therefore removed. This upgrading program began in 1981.

Unfortunately, not long after this upgrade program was completed, it was determined that the retrofitted struts had all too often suffered corrosion and metal fatigue from continued exposure to engine exhausts. An interim strut repair, which cost $100,000 per strut, was developed at Tinker AFB, Oklahoma. A complete repair would have cost $1 million.

It is worth explaining the difference between early turbojet engines and the later turbofan type. In a turbojet engine, all the air goes through the core of the engine and is compressed before it is exhausted out the back end. In a turbofan engine, a large fan is mounted in front of the core. Some of the air that is ingested still goes through the compressor and then out the back. But the majority of the air bypasses the core and goes out the back end of the engine, where it envelops the hot air coming out of the core. Today, virtually all engines are of the turbofan design, with bigger and bigger fans and more and more air bypassing the core.

The first Stratotanker converted to the KC-135E variant was delivered in March 1982. A total of 134 of the A-model tankers were eventually converted to the E-model. All of the KC-135E tankers were retired by 2009. Many of them are mothballed and sit at the USAF aircraft graveyard located at Davis-Monthan AFB, Arizona.

KC-135Q

In the late 1950s, the Central Intelligence Agency acquired a new reconnaissance aircraft from Lockheed. The design team was led by its famous chief engineer, Clarence "Kelly" Johnson. Known as the U-2, this aircraft could operate over long distances at altitudes of up to 70,000 feet. However, the U-2 had a slow cruising speed. In 1960, a U-2 flying over the Soviet Union was shot down by a surface-to-air missile (SAM).

Fig. 18-1 Photo Of Lockheed U-2

The CIA, realizing that the U-2 was no longer safe from missiles, asked Kelly Johnson to design a new reconnaissance aircraft that could fly even higher and at very high speeds. The result was the SR-71, a truly fantastic airplane that could cruise at over 80,000 feet at Mach 3, three times the speed of sound. A lot of new technology went into the SR-71, including a new fuel called hydrazine.

Fig. 18-2 Photo Of Lockheed SR-71

Some 56 KC-135A tankers were assigned to this special mission and were identified as KC-135Q tankers. The fuel initially used by the SR-71 was JP-7. It was highly corrosive, and as a result, some of the fuel tanks and fuel lines in these tankers had to be lined with ceramic anti-corrosive material. Eventually, JP-7 was found to be unusable and was no longer used. At that point, the ceramic material was removed from the 56 KC-135Q aircraft since it was no longer required.

The first Q-model tanker became available in 1962, and by 1966, 21 of the KC-135Q tankers were flying. The KC-135Q tankers, together with the SR-71 aircraft, were all based at Beale AFB, California. As SR-71 missions began to increase in the 1960s, it became apparent that more KC-135Q tankers were badly needed. This need arose from the fact that a single SR-71 mission required multiple refuelings by a number of tankers. If a single tanker missed its rendezvous with an SR-71, then the mission had to be scrubbed. Considering that each SR-71 mission was considered critical and cost hundreds of thousands of dollars, the Air Force decided that backup tankers were a priority.

After some study, SAC determined that 35 additional KC-135Q tankers were required, but that they only needed to be partially modified. Nevertheless, a total of 56 of the A-model were modified to the SQ variant. The partial modification came about because it was determined that JP-7 fuel needed to be carried only in the body fuel tanks and not in the wing tanks. When refueling an SR-71, the center of gravity of the tanker moved. Accordingly, ballast weighing 850 lb. was added to the lower nose compartment. The SQ-model tankers carried a maximum of 74,490 lb. of JP-7 fuel and a maximum of 110,000 lb. of standard JP-4 fuel, which gave the tanker its range.

The refueling of SR-71s was carried out at the extreme end of the flight envelope for both aircraft. Refueling was done at 26,000-27,000 feet, an altitude too low for the SR-71 and as high as the KC-135Q could fly when carrying a maximum fuel load. Speed for both aircraft was 355 KIAS. This was the fastest the tanker could fly and was very near the stall speed of the spy plane at those altitudes. Later flight tests showed that the tanker could actually fly safely at 375 KIAS, a better speed for the SR-71. Nearing the end of a refueling, the SR-71 would reach a point where, due to its proximity to a stall, it had to use its afterburners to maintain contact with the tanker.

It was a common procedure for SR-71s to take off from their base with only a partial fuel load. This was done to reduce stress on the landing gear tires and brakes in case of an aborted takeoff, and to ensure greater safety if one engine lost power on takeoff. Refueling an SR-71 required a great deal of skill on the part of the receiver pilot. This was clearly demonstrated one day in June 1970 when an SR-71 was being refueled over Texas. The SR-71 got too close and hit the tanker's tail stabilizer. The SR-71's nose was demolished, and its crew of two ejected safely from the aircraft. The tanker crew pinpointed where the SR-71 flight crew landed and reported that position so that the two pilots could be picked up quickly. The tanker had experienced major damage to its stabilizer but managed to return safely to its home base at Beale AFB, California.

In 1990, during a bitter political squabble between the Pentagon and Congress, the entire SR-71 program was canceled, and the aircraft were mothballed in case of future need. The 56 KC-135Q tankers were converted back to carrying JP-4 fuel only and were redesignated as KC-135T tankers.

KC-135T

As mentioned above, 56 KC-135Q tankers that had been used to refuel SR-71 reconnaissance aircraft were converted to the standard refueling configuration. They became identified as KC-135T tankers.

A second engine modification involving 440 KC-135A or KC-135T tankers took place beginning in 1979. This modification involved retrofitting the tankers with GE/SNECMA CFM56 bypass turbofan engines, each producing 22,500 pounds of thrust on takeoff, nearly double the thrust of the original J57 engines. These aircraft were then designated KC-135R. The CFM56 engines were 25% more fuel-efficient and enabled the tankers to offload 50% more fuel on long-haul missions. Furthermore, the engines increased the operational range of these tankers by 60% and were quieter on takeoff.

In June 1982 the Boeing Wichita plant rolled out the first KC-135R conversion of the A model. This engine, sometimes referred to as a CFM56 engine, was built by CFM International, a company owned 50% by General Electric and 50% by SNECMA, a French aerospace company.

Other changes included the removal of the water injection system, replacing the original Auxiliary Power Unit (APU) with a dual APU system, making the horizontal stabilizers larger, and replacing the autopilot yaw damper, which was supposed to dampen the Dutch roll. Also, a Turbofan Engine Monitoring System (TEMS) was added which recorded engine parameters during flight. Furthermore, integrated drive generators for the electrical system were added, the landing gear was strengthened and anti-skid brakes were installed.

By 1993 some 432 of the KC-135A tankers had been converted to the KC-135SR variant. Basically, two of the SR model tankers could now do the work of 3 of the A model tankers. Approximately 244 of this model are still flying today.

In 1998/99 the Air Force began a conversion of 45 KC-135R tankers to enable them to refuel U.S. Navy and Marine aircraft, as well as NATO aircraft, that use the probe and drogue system to receive fuel. To accomplish this, multi-point refueling systems were installed in these tankers. The system required the installation of a hose and drogue pod at each wingtip. The KC-135R tankers could then refuel two aircraft at the same time. This, of course, made these 45 tankers more flexible since they retained their centerline refueling booms. The design and testing of this system was carried out at Wright-Patterson AFB, Ohio.

Other Upgrades

In the years 1999-2001 a number of the KC-135 tankers received an upgrade of their flight deck instrumentation and communication equipment. This upgrade included new liquid-crystal displays, radio altimeters, autopilot, digital flight directors and other digital instruments. This upgrade was carried out under a $910 million program that was awarded to Boeing. Later, Rockwell Collins was chosen by the Air Force to introduce further upgrades. These included upgrades to compass, radar and Global Positioning System (GPS). The flight management system was integrated with an air traffic alert system.

KC-135RT– Special Operations

This model name was given to eight R-model aircraft assigned to McConnell AFB, Kansas, and operated in support of special operations missions.

Eight KC-135R aircraft were reconfigured to refuel special operations Lockheed C-130 Hercules aircraft. These missions were considerably different than the typical tanker operations of refueling bombers and fighters at high altitudes. For special operations missions the KC-135s had to fly only at night and at very low altitudes. The low altitudes were necessary in order for both the tanker and receiver to remain below land-based radar, which is line of sight. Radar beams cannot bend to follow the curvature of the Earth. Thus, an aircraft 200 miles distant from the radar source can remain below the radar beams and not show up on radar screens.

For tanker crews to operate safely at these low altitudes at night required special training, just as it did for all Special Ops personnel.

KC-135T

The KC-135T tankers were previously KC-135Q aircraft that were no longer needed due to the grounding of the SR-71 reconnaissance aircraft. These were part of the 500 aircraft that were originally retrofitted with CFM turbofan engines.

In time, the Q models underwent a second conversion, which involved an upgrade to CFM-56 turbofan engines, a new electrical generator system, a strengthened main landing gear, and anti-skid brakes. These tankers were redesignated as KC-135T aircraft.

Accordingly, in 1998, a major contract was awarded to the Boeing Company worth over $1 billion over 9 years. Boeing, in turn, awarded some of the work to its partner, the Ogden Air Logistics Center in Utah.

As of May 2018, the Air Mobility Command managed 396 KC-135 tankers. Of these, 243 are assigned to Air Force Reserve and Air National Guard units.

Conclusion

The reader can see from reading this chapter that the Air Force clearly intended to continue using the KC-135 tanker fleet for many more years. The upgrades the aircraft received made the KC-135 more useful and more valuable. However, like all aircraft, the KC-135 required more maintenance as the total flight hours of each tanker kept increasing. In the next chapter, the reader will see what the major maintenance programs accomplished.

CHAPTER 19

KC-135 MAJOR MAINTENANCE PROGRAM

Considering that the last KC-135 was delivered in 1965, and that many of them are still flying in 2025, 60 years later, it becomes obvious that these aircraft were likely to have accumulated the problems generally associated with old aircraft that have flown for many thousands of hours. Over the years such aircraft are subject to metal fatigue and corrosion, especially in the wing roots, where the wings are connected to the fuselage and where the maximum stress occurs. Serious corrosion can also develop elsewhere on an airframe.

Inspections of KC-135 tankers during routine maintenance at Tinker AFB, Oklahoma, revealed serious corrosion on the underside of the wings. As a result, a contract was let to Boeing in 1975 to replace the lower wing surfaces. Over a 13-year period, ending in 1988, a new aluminum-alloy skin was grafted to the underside of the wings. This work was carried out on 746 tankers. It included replacing the lower wing skin, which totaled 1,500 square feet per aircraft, and replacement of engine strut fittings. Each KC-135 required the installation of 564 individual parts, 32,200 steel fasteners, and 19,500 aluminum rivets. This project added 27,000 flying hours to the total life of the R and T model tankers.

A KC-135 undergoing routine maintenance was found to have deep corrosion in the belly double skin. This triggered an alarm in the Air Force, and the KC-135 tankers began to have their belly skins inspected more thoroughly. The double belly skins were corroded not only on the outside (the more likely place) but also on the inside. Removing the double skin is an extremely laborious job, especially considering that some 5,000 fasteners first had to be removed. Stringers also had to be replaced. It took 24 days for each KC-135 to have its double belly skin replaced.

In February 2017 a KC-135R flew to Tinker AFB, Oklahoma, for scheduled maintenance. A mechanic inspecting a wheel well found corrosion there. He also discovered 1.5 inches of corrosion on the keel beam, a major structural component of the aircraft that provides stiffness to the fuselage. The keel beam is 5 to 6 feet long. Fifteen KC-135R tankers were discovered to have the keel beam corrosion. Within a year the maintenance personnel at Tinker AFB were able to replace the keel beams in all 15 aircraft. All 396 KC-135 tankers remaining in the inventory at that time were thoroughly inspected for this particular corrosion problem.

At the time the KC-135 Stratotanker fleet was transferred from the Strategic Air Command to the Air Mobility Command, some of these tankers had already been retired. This took place as new KC-10 Extender tankers began to be delivered. Now the Air Force decided to extend the operational life of the remaining KC-135 aircraft. The E model tankers had their total lifetime extended to 36,000 flying hours, while the R, RT, and T models had their respective lifetimes extended to 39,000 flying hours. The Air Force estimated that these tankers already had between 12,000 and 14,000 flying hours, representing only one third of their extended lifetime. Based on that estimate, the KC-135 Stratotankers should be able to fly until 2040. In the meantime, KC-135E tankers will be slowly retired.

As to the KC-10 Extender fleet of 59 tankers, Boeing received a $216 million contract to upgrade these aircraft with new communication, navigation, surveillance, and air traffic control management systems.

Conclusion

These major maintenance programs came not a moment too soon. The tanker fleet was about to be involved in lengthy wartime operations, as related in the next chapter.

CHAPTER 20
THE TANKER FLEET AND WARTIME OPERATIONS

From 1986 onward, tanker operations changed dramatically. Except for the Vietnam War, tanker operations had been confined mainly to training and refueling aircraft during routine deployments. The Chrome Dome mission had ended in 1968. Beginning in 1986, the U.S., under the leadership of various presidents, became engaged in a series of overseas military operations. These operations were not only far from the U.S. but also had the potential of exposing tankers to enemy fighter and missile attacks. During this period of military operations, the tanker fleet truly showed how critically important it was in enabling the U.S. to project its military strength far from its shores. Some of the major operations carried out during this period are described below.

Operation El Dorado Canyon, 15 April 1986, The Libyan Air Strike

In March and April, a number of terrorist attacks were initiated against the U.S. One was a bomb set off in a discotheque in West Berlin frequented by many U.S. military personnel, injuring over 200 people. A few days later, a bomb on a TWA passenger jet, flying from Rome, Italy, to Athens, Greece, blew a hole in the side of the aircraft. Five people fell to their deaths due to cabin decompression. The crew managed to land the aircraft safely in Athens.

U.S. intelligence discovered these and other terrorist attacks at that time were engineered by Colonel Moammar Gaddafi, the president of Libya. On April 10, 1986, U.S. President Ronald Reagan decided these attacks had to be stopped and authorized the military to take action against Libya and its ruler.

The result was Operation El Dorado Canyon, a combined air strike by the Air Force and the Navy against the Libyan dictator. The raid employed a force of 24 F-111 fighter-bombers based at RAF Upper Heyford AB, England, plus Marine EA-6B Prowlers from Sixth Fleet carriers in the Mediterranean Sea, which were used to jam Libyan radars. The mission was made far more difficult because no European country would allow the F-111s to cross its airspace. As a result, the F-111s had to fly south from England to the Strait of Gibraltar and then east through the Mediterranean Sea. This routing resulted in a distance far greater than the range of the F-111s. In fact, it added 2,600 miles to the round trip.

The answer to this dilemma was a complicated utilization of tankers. KC-10 tankers from bases in Louisiana and South Carolina flew east to near the UK, where they refueled KC-135 tankers from RAF Mildenhall. The KC-135 tankers, in turn, provided the fuel to the F-111s. The F-111s had to fly a 6,400 nautical mile round-trip in order to complete their mission. Each F-111 required 8-12 in-flight refuelings. This mission could not have been carried out without the use of air tankers.

What was truly amazing about this entire operation was that few of the F-111 pilots had ever experienced an in-air refueling, and some had never even seen a tanker. Even more amazing, this refueling was carried out at night and in total radio silence! Not a single refueling was missed, and the F-111 fighters all returned safely to the UK.

The refueling plan was for KC-135s flying near Land's End at the southern tip of England to refuel KC-10s. These, in turn, refueled the F-111s all the way to Libya and back.

In addition to the F-111s flying the actual mission, other F-111s flew a decoy mission at the same time.

The First Gulf War – Operation Desert Shield & Desert Storm, 1990–91

These two major operations are sometimes referred to as the First Persian Gulf War. Operation Desert Shield was the preparation for Operation Desert Storm. Without Desert Shield, there could not have been Desert Storm. In terms of the tanker operations, it is sometimes difficult to separate the two.

Operation Desert Shield

On August 2, 1990, the dictator of Iraq, Saddam Hussein, unleashed the Iraqi Army on the neighboring country of Kuwait. The Iraqi troops quickly overcame the small Kuwaiti Army. U.S. President George H.W. Bush decided this invasion required a response. He formed a coalition of nations and created a coalition military force.

Under Operation Desert Shield, which began on August 7, 1990, a main military base was established at Jeddah, Saudi Arabia. Military assets were moved to Saudi Arabia, both to protect that nation from a possible Iraqi invasion and to prepare for the succeeding military operation, known as Desert Storm.

A main tanker base was established, also at Jeddah, Saudi Arabia. There, a huge tanker wing, the 1709th Air Refueling Wing, was created. The 1709th was composed of 76 tankers, 126 aircrews, and more than 2,000 maintenance personnel. The 1709th Tanker Wing was the largest tanker force ever assembled in one location.

As part of Operation Desert Shield, the U.S. tanker fleet was mobilized. The tankers not only made possible the transfer of fighters and transports across the Atlantic Ocean and the Mediterranean Sea to Saudi Arabia, but also B-52 bombers to Andersen Air Base on the island of Guam, located in the Marianas Islands in the Pacific.

During the initial phase of Desert Shield, there was no aerial refueling in the eastern Mediterranean Sea. The command authority quickly recognized the problem, and tankers were rushed in to fill the gap so that fighters and airlifters could reach the Middle East bases.

The tankers not only made it possible for aircraft to move quickly to Saudi Arabia, the tankers themselves, particularly the KC-10s, also acted as airlifters when needed. A tanker fleet of 300 aircraft was mobilized and dispatched to 10 separate bases in the area. An additional 100 tankers operated under the control of the USAF Europe command.

The distances that were involved in supporting Operation Desert Shield and Desert Storm were a major hurdle. As just one example, the distance from Langley AFB, Virginia to Dahran AB, Saudi Arabia, was 6,000 nautical miles. The air tankers had to refuel SAC bombers, TAC fighters and a large number of transports. In addition, KC-10s were also tasked to act as cargo aircraft in some cases. In this airlift role, seven KC-10 aircraft flew 379 missions, carried 1,111 passengers and delivered 19,905 tons of cargo.

On August 26, the first KC-135R tankers arrived at Seeb, Oman. By January 17, 1991, when Desert Storm began, 362 KC-135 and 46 KC-10 tankers were operating from 21 locations in 10 different countries.

During the period of almost six (6) months of Desert Shield, the tankers completed 4,967 sorties that totaled 19,700 flying hours. They off-loaded 28.2 million gallons of fuel to 14,588 receivers. These statistics clearly illustrate the importance of the tanker fleet presence. It should be pointed out that the KC-10 tankers also provided a majority of the airlift, especially in the beginning.

It is difficult to separate the tanker operations during Desert Shield from those during Desert Storm. During Desert Shield, the tankers made possible the non-stop flights that moved an army and all its equipment from the Continental United States (CONUS) to Saudi Arabia, a distance of 7-10,000 nautical miles, depending on the point of origin in the U.S. Much of the air refueling was often over water.

It is safe to say that without the tanker fleet the military build-up in Saudi Arabia would either have taken much longer or possibly even have been impossible. It should also be remembered that the tanker mission during the First Gulf War was flown not only by active duty Air Force crews but also by Reserve and National Guard crews.

Operation Desert Storm

In contrast to Desert Shield, much of the air refueling during Desert Storm involved wartime missions near the actual combat zone. Fighters and fighter-bombers going into the battle area had to be refueled in order to be able to return safely to their base in Saudi Arabia.

On January 17, 1991, Operation Desert Storm began with the bombing of strategic targets in Iraq. The KC-10 and KC-135 tankers from the 1709th Wing began fueling missions that took them over enemy territory. Many of the tanker crews flying these missions had never flown wartime missions. Early on that day, some tankers were refueling F-15 fighters at 3,000 feet just before the fighters initiated their attacks. In the first days of Operation Desert Storm, hundreds of tankers were flying sorties simultaneously. On the first day alone, the tankers flew 103 refueling missions. This level of activity required tremendous planning and coordinating effort.

At the peak of Desert Storm, 81% of the KC-10 fleet and 44% of the much larger KC-135 fleet were involved in supporting the operation.

Fig. 20-2 Map of Area of Desert Storm

The air campaign against Iraq in 1991 had the tanker fleet operating at its peak operational ability. More than 300 tankers were deployed, using 122 separate air refueling tracks. Desert Storm lasted a total of 43 days.

For the total of both Desert Shield and Desert Storm operations, the 1709th Refueling Wing flew 51,000 refueling sorties and transferred 125 million gallons of fuel.

Once Operation Desert Storm began, the total tanker force included not only U.S. Air Force tankers but also Navy and Marine tankers, plus a few tankers from coalition countries such as the UK, France, Canada, and Saudi Arabia. The U.S. Air Force also called up to active duty some 200 personnel from the Reserves and the National Guard to help support the huge effort required.

As part of this wartime operation, a total of 160 refueling tracks were set up. However, 19 of these tracks accounted for 80% of all refueling conducted during Operation Desert Storm. During the most intense parts of Operation Desert Storm, as many as 300 refueling sorties were scheduled in 24 hours. This level of activity caused air traffic problems and sometimes even limited the number of sorties flown.

During the 43 days of Operation Desert Storm, the U.S. and coalition tanker fleet flew 15,434 sorties and offloaded 10.2 million gallons of fuel. The net result was that the concept of an "air bridge" was fully developed and implemented.

The tanker crews flying the missions during Desert Storm encountered various obstacles and yet always managed to complete their assigned tasks. The descriptions below indicate the caliber of flying these crews performed.

Brig. Gen. Kenneth Keller, USAF (Ret.), commanded the entire air refueling operation for both Desert Shield and Desert Storm. He remembers that, once Desert Storm began, the tankers were not just very busy carrying out pre-planned missions, but also had to perform additional refueling as emergencies developed. During the entire First Persian Gulf War, not a single refueling rendezvous was missed. That alone speaks volumes as to the professionalism of the tanker crews as well as all the supporting personnel.

Mission Details

On the night Desert Storm began, Captain David Horton was the Aircraft Commander (A/C) of a KC-135R tanker. The mission was to refuel an F-117 stealth fighter. The fighter was returning from its mission and, due to foul weather, had missed its assigned refueling hook-up. Moreover, the F-117 was still carrying a very heavy bomb that the pilot had been unable to drop due to the weather. In addition, the fighter was very low on fuel. The F-117 contacted an Airborne Warning and Control (AWACS) aircraft flying high overhead and received directions to an alternate landing field. Captain Horton heard the radio call and realized the F-117 would be flying near him on the way to the alternate field. The KC-135R was flying in clouds, so Horton began climbing to reach clear air, eventually reaching clear sky at 27,000 feet.

The two aircraft managed to rendezvous, but the F-117 could not stay at that altitude because of the very heavy unused bomb it still carried. Meanwhile, the fuel tanks of the F-117 were almost empty.

The boom operator managed to connect to the F-117, but that aircraft was having great difficulty remaining at that altitude. Captain Horton, realizing the predicament of the F-117 pilot, did what is called a "toboggan" maneuver. He put his aircraft into a shallow descent, thus picking up enough speed for the fighter to continue accepting fuel. After unhooking from the tanker, the F-117 pilot thanked the tanker crew and made it safely back to his base.

In another case, Major Diane Byrne, the A/C of a KC-10, was assigned to a fueling track over Iraq. Once on the track, she was directed to descend from the typical refueling altitude of 25,000 feet down to 12,000 feet and locate a C-130 needing fuel. Because of the frequent presence of Iraqi anti-aircraft fire, tanker hook-ups had to be done in total radio silence.

Major Byrne could not locate the C-130, so she put out a brief radio signal. This finally made it possible for the two aircraft to rendezvous, and the tanker began pumping 30,000 pounds of fuel into the C-130. Afterwards, Major Byrne acknowledged that her main concern had been the possibility of a mid-air collision with another aircraft because of the low altitude at which the refueling was carried out.

Captain David Gordon, USAF (Ret.), remembers that a typical wartime mission near Iraq for a KC-10 tanker might involve a 1.5-hour flight to a general refueling area, orbiting for 4 hours, and then flying back to home base for 1.5 hours, for a total flight time of 7 hours. Capt. Gordon also remembers one mission in the Middle East that ran much longer. The tanker was under the control of a USAF AWACS aircraft flying in the general area. His tanker was directed to fly to another area and orbit there before returning to base. The KC-10 took on some 120,000 pounds of fuel from a KC-135 tanker, remained in the new area for 4 additional hours, refueled a number of aircraft, and finally landed back at its base after a total of 12 flight hours.

Lt. Col. Marc Russick, USAF (Ret.), remembers flying a KC-10 on a very long non-stop mission from Travis AFB, California, to Al Dhafra, United Arab Emirates, with cargo and passengers that took 17 hours. The flight was denied landing rights in both Egypt and Saudi Arabia for diplomatic reasons. As a result, the KC-10 refueled from another KC-10 in order to complete the mission.

Lt. Col. Michael Moore, USAF (Ret.), also flew the KC-10 tankers over Iraq. He remembers his longest flight as one from Washington, DC, to Seoul, South Korea. That flight took 17.5 hours and one air refueling. The longest basic sortie he flew lasted 11.5 hours and was flown with no extra crew. Such flights required the tanker to refuel from another tanker a couple of times while also, in turn, refueling other aircraft.

Col. Mark E. Smith, USAF (Ret.), was a fighter pilot who flew F-4 and F-15E fighters and served from 1974 to 2000. In an interview with this writer, he described what a night operation was like during Operation Desert Storm when he flew as part of the 1st Fighter Wing. A large number of F-15E Eagle fighters departed their base at Dahran, Saudi Arabia, in total darkness. The distance from Dahran, Saudi Arabia, to Baghdad, Iraq, is roughly 600 nautical miles. En route to their target in Iraq, the fighters refueled once. This was accomplished without any radio communications whatsoever, on a pitch-black night with no moon shining. The fighter pilots used such navigation systems as TACAN and radar. The tanker boom operators assisted by using lights attached to the rear of the tankers.

The fighters normally refueled a second time when returning from their target. If the mission became longer due to in-flight changes, then a third refueling would be necessary before the fighters recovered at their home base.

From the above description, it is apparent that the tanker fleet was very busy on a 24-hour basis. Not only were the tankers flying scheduled missions, but they also remained on station, available to assist in any eventuality that might arise. During wartime, tanker missions were at times very lengthy and could involve changes from the original flight plan as situations demanded.

After the First Gulf War ended, top military commanders all agreed that tankers were the one military asset that made it possible to fight and win the entire war in such a short period of time.

The Saga of Whale 50

During the First Gulf War, tanker operations in that theater were intense. One such mission came very close to disaster. A KC-135 was launched from Jeddah AB, Saudi Arabia, for a night refueling mission. Its radio call sign was "Whale 50". Another tanker was dispatched immediately behind Whale 50. The second tanker had orders to pass Whale 50 sometime after takeoff. Evidently, the layout of Jeddah AB plus the number of tankers located on the ramp there prevented the second aircraft from departing first.

After Whale 50 reached an altitude of 25,000 feet, the second tanker passed it at 27,000 feet on its left. A few seconds later, Whale 50 went into a violent left bank of 90 degrees, followed immediately by a second equally violent bank of 90 degrees to the right. The plane's gauges showed that all power had been lost on the two turbofan engines mounted under the left wing. The boom operator went to visually check what had happened. He reported that both left-hand engines had departed the wing and were gone.

Fig. 20-3 Photo of KC-135 Minus 2 Left Engines

The A/C immediately declared an emergency and alerted Jeddah AB that they were turning around and returning to Jeddah. Both pilots were straining to keep Whale 50 level. This was not easy since, with the two left-hand engines gone, there was much more weight on the right side. The A/C decided that the two pilots would alternate at the controls every few minutes so that each pilot could rest briefly while the other one wrestled with the control wheel.

After turning back, the crew of Whale 50 dumped most of its fuel in preparation for what they already knew would be a very challenging approach and landing. With the two left engines gone, the aircraft had lost most of its hydraulic pressure. This meant that the landing gear would have to be lowered manually. It also meant that once the aircraft landed and the brakes were applied, there was only enough hydraulic pressure for one extended brake application. The brakes could not be pumped. In addition, using engine reverse thrust on the remaining right-hand engines would create an asymmetrical condition which would almost certainly force the aircraft to exit the runway with likely dire consequences.

The crew made its approach to Jeddah AB, touched down, and after a long roll-out with the brakes full on and smoking, managed to stop the aircraft before running out of runway.

An investigation was immediately launched to determine what had caused such a weird and potentially fatal situation. A USAF accident investigator was quickly sent to Saudi Arabia. Luckily, the two engines that had separated from the aircraft were found by wandering Bedouins, and, after being notified, the Air Force brought them back from the desert. The investigation concluded that Whale 50 had encountered turbulence created by the second KC-135 flying only 1,000 feet higher. This turbulence caused Whale 50 to initially roll violently to one side. This violent roll is what caused the engines to separate from the wing. The Boeing Company designed the KC-135 so that if excessive stress was applied to a wing, the two engines on that side would separate before the wing could suffer catastrophic failure.

Iraqi Operations – 1992-2003

In the aftermath of Desert Storm, a number of operations took place involving Iraq and its military, as shown below.

Operation Southern Watch – Aug. 1992 to March 2003

In 1992, the U.S. Air Force began an operation called "Southern Watch" that required Air Force fighters to fly over Iraq and maintain a no-fly zone over the south-central portion of that country. Operation Southern Watch was in direct response to the Iraqi Air Force attacking

Shiite Muslim areas in southern Iraq. This operation was conducted continuously from the end of the first Persian Gulf War until the beginning of the second Gulf War. The direct result of Operation Southern Watch was that a no-fly zone was created in Iraq south of the 33rd parallel and was maintained for almost eleven years.

Operation Provide Comfort – April to July 1991

Operation Provide Comfort was basically a humanitarian operation that delivered fuel and cargo to the Kurdish Region of Iraq, which had been under siege by the Iraqi dictator, Saddam Hussein. During this short period of time, KC-135 tankers flew a total of 905 sorties and delivered 4.5 million gallons of fuel to cargo aircraft involved in the operation.

Operation Restore Hope – December 1992 to May 1993

This operation supported the dispatching of 25,000 troops to Somalia to bring food to that country. Here, the principle of an "air bridge" was demonstrated again. The distance of over 7,550 miles from the U.S. would have prevented the quick movement of food, troops, and supplies if there had not been an "air bridge". The net result was saving thousands of people from starvation. Air refueling made this operation possible.

Operation Vigilant Warrior

In late 1994, Saddam Hussein moved Iraqi Republican Guard forces to the border with Kuwait. This move alarmed the U.S. government, which responded by rushing troops to the area. As a result, the Iraqi forces were withdrawn. Once again, tankers were a critical part of the operation.

Operation Strike

In 1996, Iraqi forces invaded the Kurdish regions of northern Iraq. American forces were rushed to the region in order to protect the

Kurdish population. Tankers supported this operation. In one part of this operation, B-52 bombers from Barksdale AFB, Louisiana, flew round trips of 13,683 nautical miles to launch cruise missiles against targets in Iraq. These were 34-hour flights supported by 15 KC-10 and KC-135 tankers which offloaded 1.4 million pounds of fuel.

Operation Desert Fox

In November 1998, at the order of then-President Bill Clinton, the U.S. military carried out a four-day strike against targets all over Iraq. Once again, USAF tankers were involved.

Operation Northern Watch – January 1997 to September 2002

Operation Northern Watch was launched to enforce the no-fly zone over Iraq north of the 33rd parallel.

The Air Mobility Command fleet of KC-135 and KC-10 tankers was instrumental in two ways during these Iraqi operations. First, it enabled USAF fighters such as the F-15 and airlifters to reach Saudi Arabian bases flying non-stop from the U.S. by en route refueling. Second, some of the tankers based in the Middle East supported the fighter missions and provided a level of security not otherwise available. Furthermore, the tanker fleet enabled the U.S. presence to continue without interruption until 2003.

The Kosovo War – 1998-99

The Kosovo War was fought entirely within what had been the country of Yugoslavia. The war involved Serbia and Montenegro, whose forces attacked a rebel army in the province of Kosovo.

On February 28, 1998, war erupted between the two sides. NATO voted to intervene. It did so by mounting air strikes against the invading forces of Serbia and Montenegro, beginning in March 1999.

The air war, which was mainly carried out by the U.S. Air Forces in Europe (USAFE), required the mobilization of fighters, bombers, and tankers to Europe. The USAF provided 90% of the total tanker force used in this operation. This tanker force was made up of 112 active duty and 63 reserve tankers. The tanker force flew over 5,000 sorties in support of 24,000 combat and combat-related sorties. In total, the KC-10 and KC-135 tanker force supplied 250 million pounds of fuel. Critical to this entire effort was the maintenance that kept the tankers flying.

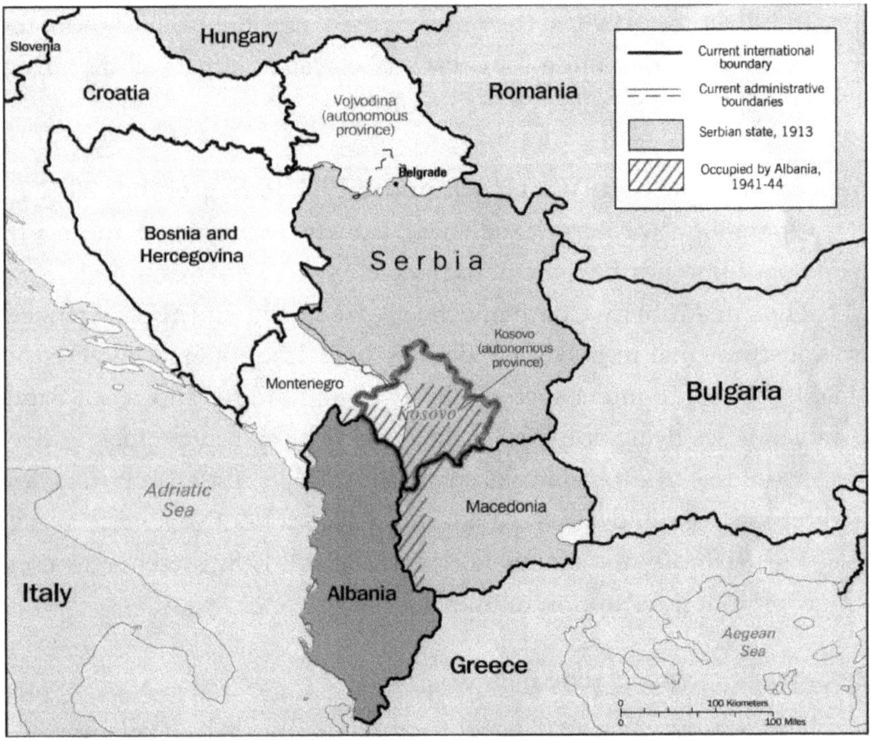

Fig. 20-4 Map of Kosovo and Surrounding Area

The tanker missions were naturally very closely coordinated with the battle sorties of NATO aircraft, most of which were members of the U.S. Air Force and Navy. Two examples show the close coordination that made the entire effort a success.

In one instance, a Royal Air Force (RAF) KC-135 from Mendenhall AB, UK, came within 70 miles of two Serb MiG-29 fighters. At jet speeds, this is not a great distance. Two F-15C fighters engaged and shot down the MiG aircraft.

In an even more dramatic instance, an Air Force F-117 was shot down over Serbia. The pilot successfully ejected, parachuted to the ground safely, evaded capture, and radioed for rescue. A force of over 20 tankers maintained a search and rescue effort for over six hours until the pilot was safely extracted.

At the beginning of this war, U.S. Air Force B-2 bombers flew from their home base at Whiteman AFB, Missouri, to Kosovo, dropped their conventional bomb loads, and flew back to Whiteman AFB, all non-stop. The one-way distance was approximately 5,400 nautical miles. This meant a 31-hour flight that required four air refuelings. It was truly a dramatic demonstration of the USAF's Global Reach strategy, only made possible by the use of air tankers.

As it happens, the day the air war in Kosovo began, this writer was scheduled to fly on a Turkish Airlines flight from London Heathrow Airport to Istanbul, Turkey. The aircraft was very late arriving from Istanbul. After it was serviced and everyone was aboard, we sat on the tarmac with the door open for over two hours without any explanation as to why that was happening. Finally, the flight was cleared by the control tower to start engines. The captain was obviously in a big hurry due to the long delays. The door was closed, the engines were started, and we taxied at a very fast rate past five other aircraft on a parallel taxiway, turned onto the runway, and without stopping took off. It was pretty obvious the captain was afraid of losing his takeoff slot. We followed a route far north of the normal route and finally arrived in Istanbul many hours late. We later learned that the delays were caused by European air traffic control agencies rerouting many flights due to the fear of mid-air collisions with NATO military aircraft.

A major problem arose for the tankers between March 24 and June 8. The tanker force grew from 55 to 175 aircraft. This number of tankers created the issue of where to base them. USAFE had surveyed many former fighter bases in Europe, but most could not serve as bases for such large aircraft as the tankers. Eventually, the tankers were based in such far-off locations as Budapest, Hungary, and a French air base at Mont-de-Marsan. There were 12 tanker bases stretching from England to Eastern Europe and the Mediterranean. The RAF base at Mildenhall hosted 34 tankers, and the base at Morón, Spain, accommodated another 38.

During this war, the tanker force continually increased in size. This was a function of tankers being dispersed to distant bases, which resulted in them carrying less fuel per delivery due to transit times of up to three hours in each direction. Similarly, some fighters based as far away as the UK required much more fuel to complete their missions over Kosovo. Another problem arose from some countries not allowing NATO aircraft to transit their airspace, which further increased the overall fuel required.

The KC-135's mission-capable rate was 78 percent, while the much newer KC-10's rate was 88 percent. This was a testament to the quality of maintenance carried out during the air war.

At the peak of the Kosovo air war, 40% of the tanker fleet was in use, and 80% of tanker crews were employed. These numbers showed the USAF and Air Mobility Command the need to generate as much efficiency as possible in any future large-scale operation.

One example of tanker use was the following: Fully fueled KC-10s flew a reliability orbit throughout the Kosovo air war. They had no assigned receivers. They were there to handle refueling when a tanker had a maintenance problem and had to cancel its scheduled flight, or to refuel fighters that needed extra fuel. Some tankers were kept on ground alert in case they were also needed.

One outgrowth of the Kosovo War was the requirement for a much improved Command and Control system for scheduling and dispatching tankers. The air war in Kosovo began fairly small and then expanded quickly. The personnel in charge of tanker force operations were inundated with questions and urgent requests for specific sorties. Initially, this placed a very large responsibility and a high degree of stress on the small group of personnel assigned to direct the tanker force. Eventually, a colonel with a great deal of experience in tanker operations arrived on the scene, and tanker operations began to function much more smoothly.

The initial operational difficulties experienced at the beginning of the Kosovo War were a valuable lesson for Air Mobility Command, enabling it to prepare for large-scale future tanker operations.

The entire air war lasted only 78 days, and remarkably, not a single airman lost his life. To put the role of the tanker force in the Kosovo War in perspective, the war was won entirely by the efforts of the NATO air forces. These air assets were able to carry out their missions only because of the presence and efforts of the tanker fleet. Without those tankers, there would not have been any air war to speak of.

Operation Noble Eagle - September 11, 2001

Operation Noble Eagle was in response to the terrorist attacks on the World Trade Center in New York City and the Pentagon in the District of Columbia.

On 9/11, as the terrorist attacks on the World Trade Center in New York City were being carried out, the North American Air Defense Command (NORAD) launched two F-15 fighters from Otis AFB on Cape Cod. They were launched just as the second World Trade Center tower was being attacked. In support of these two fighters, a KC-135 was launched from Bangor, Maine. The tanker set up an orbit at 20,000 feet over John F. Kennedy International Airport on Long Island, New York. The F-15 aircraft took turns refueling from the tanker. Soon, more F-15 fighters arrived on the scene, and they too took turns refueling from the same KC-135. Later, a KC-10 from McGuire AFB, New Jersey, replaced the KC-135.

From that small beginning, the Operation grew until it included National Guard, Reserve, and active duty units. The basic purpose of Operation Noble Eagle was the protection of the entire U.S. from any additional terrorist attack. To that end, fighter aircraft and supporting tankers were dispatched to a great many locations, based in part on available intelligence. The Operation lasted until December 2009, with an F-15 sortie from Homestead AFB, Florida. This sortie was the 55,000th accident-free sortie, a notable accomplishment. During this entire period, USAF aircraft were operating 24 hours a day, seven days a week.

Operation Enduring Freedom – 2001-2021

In 2001, President George W. Bush made the critical decision that terrorist groups such as Al Qaeda, which operated out of Afghanistan, had to be stopped from carrying out further instances of terrorism. In this particular case, that meant deploying troops thousands of miles to that distant country. The military operation in Afghanistan eventually became the longest war in U.S. history. Called Operation Enduring Freedom (OEF), this military operation began on October 7, 2001. This operation continued until the end of August 2021, an astounding 20 years later. Because Operation Enduring Freedom lasted so long, it is very difficult to present a historical narrative in just a few paragraphs. In fact, as of early 2023, no summary of tanker operations over that 20-year period has appeared as yet. Nevertheless, there is some information available that relates to the first few years of OEF.

Fig. 20-5 Map of Afghanistan and Surrounding Area

Actually, the Air Mobility Command (AMC) began widespread operations prior to the official start date of OEF. AMC had 378 airlifters and tankers in their active duty fleet. Of these, 94 were awaiting repairs. AMC maintenance crews jumped into action, and in two days, 55 of these aircraft were back in service.

During the previous weeks, AMC was already airlifting troops and materiel to Afghanistan. This initial effort required the support of the tanker force, without which the airlifters would not have been able to fly nonstop to Afghanistan.

From the beginning of OIF, the USAF realized that it would have to employ the concept of an "air bridge" if it was going to have any real effect on events in Afghanistan. One of the first tanker units deployed was the one at Grand Forks AFB, North Dakota. Other units deployed shortly thereafter. Within 30 hours after the start of OEF, tankers were on their way to Guam to support B-52 and B-1B bombers being deployed overseas, then supporting them pre- and post-strike. Other tankers deployed to such far-off locations as Japan and Diego Garcia, a little-known island in the Indian Ocean. The 376th Expeditionary Wing was set up at Manas, Kyrgyzstan.

The KC-135 tankers bore the burden in the beginning stages of OEF. Then KC-10 tankers were deployed to the 380th Air Expeditionary Wing in Southwest Asia.

Maintaining troops in Afghanistan required continuing tanker support to refuel fighters, bombers, and transports. Tankers operated from such diverse locations as Karshi-Khanabad in Uzbekistan and Manas AB in Kyrgyzstan, both of which had formerly been air bases of the Soviet Union before the collapse of that regime. Tankers also operated from other bases, some of which were in Europe. To reach the base in Afghanistan, which was Kandahar Airport located near Kabul, the capital, tankers had to fly up to 7,000 nautical miles. From this, the reader should be able to have some understanding of the very long distances involved and the associated difficulties that crews might encounter. For example, for a tanker operating 7,000 nm non-stop, that meant a non-stop flight of approximately 14 hours.

In 2006, the tankers off-loaded 871 million pounds of fuel to 42,083 receivers. In 2007, they off-loaded 947 million pounds of fuel to 79,798 receivers. Then, in 2008, the tankers off-loaded 1.1 billion pounds of fuel to 86,288 receivers. In 2009, they off-loaded 1.07 billion pounds of fuel to 82,095 receivers, while in 2010, they off-loaded 1.05 billion pounds of fuel to 82,063 receivers. Then, in 2011, the tankers off-loaded more than 875 million pounds of fuel to 68,203 receivers.

In the first six years of OEF, the tankers off-loaded 5.9 billion pounds of fuel to 441,070 receivers. These numbers are certainly a testament to what the tanker fleet was able to accomplish.

As an example of what the tankers made possible, B-2 bombers from Whiteman AFB, Missouri, flew to Afghanistan and back to their home base non-stop, requiring six air refuelings to complete their mission.

A large concentration of tankers operated over Southwest Pakistan. These tankers were based in Diego Garcia, Oman, and Bahrain. Bombers from various bases, including Diego Garcia, refueled in this area before entering Afghan airspace to carry out their missions.

After a while, the statistics lose their impact. The bottom line is that over the 20-year duration of OEF, the tankers made possible not only thousands of air strikes but also constituted the "air bridge" that delivered and supported troops in Afghanistan. At one point in this war, the number of troops deployed reached 100,000. The reader should not ignore the simple fact that Afghanistan is a country halfway around the world from the U.S.

The constant availability of tankers resulted in some very long-range military flights that were truly amazing. One Air Force F-15E fighter on a combat mission was airborne for 15.5 hours. A B-2 combat mission set a record when it remained aloft for 44 hours. B-1 and B-2 bombers arrived carrying precision weapons and then had to orbit over the battleground for hours while waiting for target assignments.

Air Force tankers also provided over 80% of the fuel used by Navy aircraft flying combat missions. It was not unusual for Navy pilots to be refueled six times on missions lasting from seven to ten hours.

None of these flights would have been possible without the continuing presence of tankers.

Operation Iraqi Freedom (OIF) – 2003

This operation is sometimes referred to as the Second Gulf War. It was conducted by a coalition of nations led by the U.S. The operation involved the invasion of Iraq and the toppling of the Iraqi government led by Saddam Hussein. The war began on March 19, 2003, and lasted just 43 days. A total of about 177,000 coalition troops were involved, with 130,000 from the U.S. The coalition withdrew from Iraq in 2003.

U.S. troops were initially brought to Kuwait and from there invaded Iraq. A total of 149 KC-135 and 33 KC-10 tankers initially participated in that operation. The tankers flew 6,193 sorties and off-loaded 376,391 pounds of fuel.

Between March 19 and April 18,2003, there occurred a 720-hour period when the military executed a major and very concentrated effort. This was the buildup of troops and equipment for the invasion of Iraq. During that period, 1,800 aircraft flew 41,404 sorties. It was a truly amazing effort.

Without the presence of the tankers, it would have required much more time to accomplish the safe movement of these thousands of troops and their equipment.

Operation Iraqi Freedom ended with the capture, trial, and execution of Saddam Hussein.

It appears that it was during OIF that an acronym that specifically referred to the tanker fleet was created. The acronym was "NKAWTG". The acronym stands for "NOBODY KICKS ASS WITHOUT TANKER GAS". While the saying may seem rather crude to the reader, it nevertheless demonstrates, in the clearest possible terms, how military operations far from the Continental United States cannot take place without aerial refueling. OIF was merely one more instance of the necessity for aerial refueling if U.S. diplomatic and military strategy is to be carried out.

Conclusion

The above descriptions simply reinforce the fact that tanker operations completely changed military aircraft operations from what they had been previously.

The USAF tanker fleet was about to receive some assistance, but not without a fair amount of controversy, as we shall see in the next chapter.

CHAPTER 21
BOEING, THE KC-46 AND THE PENTAGON

The history of the KC-46 tanker dates back to approximately 2000. The U.S. Air Force was concerned about the age of its tanker fleet. Specifically, the backbone of the fleet, the KC-135, by then approximately 50 years old, was reaching the point where it needed to be replaced. Maintenance on those aircraft was becoming ever more costly, and so were spare parts. Accordingly, the USAF planners began searching for a replacement. From the beginning, based on USAF experience with the wide-body KC-10, they knew that the best replacement was another wide-body. Such an aircraft would carry more fuel over longer ranges and, additionally, better serve the role of airlifter.

Initially, in late 2003, the Air Force proposed replacing 100 of the oldest KC-135E tankers with a tanker version of the Boeing 767 wide-body passenger transport, designated the KC-46. Competing with Boeing was a wide-body Airbus A330 which would be built by Northrop Grumman in the U.S. The Air Force proposed purchasing 80 of the KC-46 and leasing 20 more. At this point a furor erupted, led by Senator John McCain of Arizona. McCain, and an organization called the National Legal and Policy Center, began an investigation of the proposed deal. They went to the Pentagon Inspector General and the DOD Criminal Investigative Service, who investigated further. The net result was the uncovering of a conflict of interest on the part of a Pentagon high-level official. This official, a former Boeing executive, working together with two Boeing executives, had managed to swing the deal to Boeing. As a result of the investigation, the contract was canceled by the Pentagon.

In 2006, the Pentagon issued a new request for proposals for a tanker identified as the KC-X. This time, Boeing proposed a version of the Boeing 777 while Airbus, together with Northrop Grumman, offered a version of the Airbus A350. After more controversy, this proposal was also rejected in November 2008. Once again, the two companies submitted bids. In February 2011, the Defense Department announced the award of a contract for 179 KC-46 tankers to Boeing. The contract also called for an initial group of 18 aircraft, including test aircraft, to be delivered by 2017. The contract further stipulated a maximum of 15 tankers to be delivered each year. This was a fixed-price contract worth $4.9 billion, and it stipulated that Boeing was responsible for any financial penalties resulting from technical problems that arose once the KC-46 entered USAF service. The aircraft then became officially known as the KC-46A Pegasus. The Air Force intended to gradually retire approximately one-third of the remaining KC-135 fleet as the new KC-46 tankers became operational.

Looking back at these events, it seems quite clear that both Boeing and the Air Force were remiss in monitoring what some of their personnel were actually doing. The Pentagon should definitely have had an inkling that one of its top employees, especially one who originally came from Boeing, might be doing something illegal. The fact is that this employee should have been prevented from participating in any way in the acquisition process of any aircraft from Boeing. By the same token, someone at Boeing should have been aware that top executives were involved in negotiations with the Pentagon in a way that amounted to a conflict of interest.

Conclusion

The final result was that both the Pentagon and Boeing appeared guilty of unethical and possibly illegal conduct, as well as being incompetent in their business affairs. True, the Air Force finally did procure a new tanker, which it desperately needed. But that tanker program proved to be years late and full of serious problems. (See Chapter 23). What made things more difficult for the Air Force was the fact that the KC-135 fleet was showing its age, and its maintenance was becoming more and more expensive. (See Chapter 19). Nevertheless, the contract was signed and executed, and Boeing began the manufacture of the first batch of KC-46 Pegasus tankers.

Ultimately, after some years of indecision, the Air Force was finally getting its new tanker, the KC-46 Pegasus, as described in some detail in the next chapter.

CHAPTER 22
THE BOEING KC-46A PEGASUS TANKER

The KC-46 tanker is based on the Boeing 767-200ERF (Extended Range Freighter), which in turn is based on the Boeing 767 passenger transport. It is a wide-body aircraft and thus is able to carry more fuel than the KC-135 that it is supposed to replace. The first flight of the original B-767 was in 1981, while the KC-46 first flew in 2013, or 34 years later. This means that the basic airframe design is really quite old. However, a number of changes were introduced in the KC-46 model. The KC-46, while based on the B-767-200ERF airframe, is not identical. The KC-46 is based on a B-767-2C, which was derived from the B-767-200ER. The fuselage is the same. The -2C had a strengthened main deck floor, a cargo door, several freighter features, strengthened B-767-300ER wings, B-767-400ER horizontal stabilizers, B-787-based flight deck displays, auxiliary body fuel tanks for greater fuel capacity, and extra plumbing plus 50 miles of wiring required for the refueling mission systems. From this list, the reader can see how Boeing pieced together various parts of different 767 models to create the KC-46.

Fig. 22-1 KC-46A Refueling C-17 Airlifter

The reader should not be misled into thinking that because the basic airframe design is quite old, the KC-46 is simply an old aircraft design. The KC-46 embodies a number of new technologies which make it an improvement over previous tankers, including the KC-10. The table below gives the reader a comparison of the Boeing 767-200ERF, the KC-46A, and the KC-10.

	B-767-200ERF	KC-46	KC-10
Wingspan	165’	157’ 8”	165’
Max to Weight	395,000 lbs.	415,000 lbs.	590,000 lbs.

A comparison of these numbers clearly shows that the KC-10 was a bigger and heavier aircraft than the KC-46. But, as explained below, the KC-46 featured a number of innovations and improvements.

Specifications

FUSELAGE LENGTH 159 FT. 2 IN.

FUSELAGE WIDTH FT. 1 IN.

HEIGHT 52 FT. 10 IN.

WINGSPAN 157 FT.8 IN. with 34 DEGREE SWEEPBACK
MAX. GROSS T/O WEIGHT 415,000 LBS.
MAX FUEL CARRIED 212,300 LBS.
POWER PLANTS 2-PRATT & WHITNEY PW4062 TURBOFAN ENGINES with 62,000 LBS. THRUST EACH. Electric power provided by 120 KVA generators.

Performance
CRUISE SPEED MACH 0.80
SERVICE CEILING 40,000 FT.
RANGE 7,350 MILES
CARGO CAPACITY - 18 PALLETS
MAX. CARGO WEIGHT - 65,000 LBS.
Note that the KC-46 is considerably smaller than the KC-10 and carries a smaller fuel load.

Mission Capabilities

When the USAF developed the specifications for its new tanker, it based them on what it saw as the missions of such a tanker. These are shown below.

The KC-46 was designed to perform six (6) primary missions. These are:

1. Nuclear operations support
2. Global strike support
3. Air bridge support
4. Aircraft deployment
5. Theater support
6. Special operations support.

The Pegasus also has the capability to support the following secondary missions:

1. Airlift

2. Medical evacuations
3. Emergency aerial refueling
4. Air sampling
5. Support of combat search and rescue.

Thus, the various features identified below are a direct result of the primary and secondary missions shown above.

Additional Features

The KC-46 has a number of features that are important, as detailed below.

- Electromagnetic pulse hardening protection
- Chemical/biological operations
- On-Board Inert Gas Generation System – This system converts oxygen in the fuel tanks to inert nitrogen to prevent an explosion if the aircraft is hit by lightning or enemy fire
- Fuel tank ballistic protection
- Cockpit armor
- Large aircraft infrared countermeasures
- Radar warning receiver
- Tactical situational awareness system
- Radio frequency warning
- Night vision lighting

Exterior Lighting

- Nose gear taxi light
- Belly position lights
- Fuselage lights – both sides
- Wing-mounted landing lights – both sides
- Inner engine pylon lights – both sides
- Belly strobe
- Dorsal strobe

- Aft underwing lights – both sides
- Tip of fuselage light
- Wingtip strobe and marker lights – both sides
- Vertical stabilizer refueling floodlight
- Aft tip of fuselage light.

The KC-46 has fuel tanks located in the wings plus four (4) body fuel tanks located in the lower lobe below the cabin floor. The body fuel tanks are manufactured by the Cobham Company of Great Britain. Each tank measures 10 feet wide, 5 feet high, and 5 feet deep. These tanks pump fuel to the center fuel tank.

The KC-46 has a centerline boom for refueling large aircraft such as bombers, airlifters, reconnaissance aircraft, and command and control aircraft, plus other tankers. The Pegasus is also equipped with three (3) drogues. There is a centerline drogue plus two (2) wingtip drogues which are carried in pods. The boom is an upgrade of the one used on the KC-10 and is remotely operated. The drogues are the latest version from the Cobham Company of Great Britain, the same company that pioneered the drogue system.

The three drogues can be deployed at 180 to 325 knots indicated airspeed (KIAS). The drogues all have a flow rate of 400 gallons per minute, and all three have a trailing cable length of 75 feet.

The centerline boom and the two wingtip drogues can be deployed and used simultaneously.

The KC-46 saw a major change relating to the boom operator position, which was moved from the belly of the aircraft to just behind the flight deck on the right side of the fuselage. In addition, a second boom operator station was introduced next to the first. The boom operator operates the boom remotely using a system of electronic controls and a remote visual system.

The purpose of the second boom operator station has never been adequately explained, although one USAF source did say that the second station was for a refueling instructor. However, this author asked both the Boeing Company and the USAF Air Mobility Command the real reasons for the following innovations.

1. The Boom Operator was moved from the previously used belly position to a location just behind the Flight Deck. This relocation necessitated the creation of a complex and expensive Remote Boom Operating System, which subsequently introduced additional problems and was unpopular with many Boom Operators.
2. The installation of a second, sophisticated, and expensive Boom Operator console adjacent to the first.

To date, no answer has been found by this author.

The two issues detailed above resulted in significant financial costs to Boeing and caused long delays in the delivery of KC-46 tankers to the USAF.

Both Boom Operator stations use fly-by-wire technology but operate independently. This suggests that a KC-46 can simultaneously refuel a bomber using the refueling boom and two fighters using the two wingtip hose and drogue systems. However, this has never been clarified. The two refueling operators each have their own displays and controls and are situated next to each other. The two refueling systems use cameras and sensors to enable the operators to carry out their individual refueling duties. This change in the location and number of refueling operators proved to be not only controversial but also problematic, delaying the formal acceptance of the KC-46 by the Air Force. (See Chapter 23 for more information on the KC-46's problems.)

The KC-46 can not only refuel other aircraft but also be refueled by other tankers through the refueling receptacle located on the aircraft's roof, just behind the flight deck. Refueling the KC-46 through this receptacle can occur at a rate of up to 1,200 gallons per minute.

Additional Innovations

Flight Deck

The KC-46 has a flight deck featuring fly-by-wire flight controls. It also has four 15.1-inch liquid crystal cockpit displays similar to those introduced on the Boeing 787 Dreamliner wide-body passenger transport. Rockwell Collins supplied a signal data concentrator (SDCN)network that translates and routes data between avionics systems. SDCN is an avionics full-duplex, switched-Ethernet backbone based on the ARINC 664 specification and is used for mission integration. The flight control yokes are the standard design used for decades by Boeing.

Communication Systems

Boeing also developed a new radio antenna integrated into the skin of the aircraft. In addition, the Cobham Company supplied a two-element VHF/UHF antenna for terrestrial air-to-air and air-to-ground communications, as well as a Satcom UHF antenna for near-hemispheric coverage and a high-gain Satcom antenna.

Oxygen Systems

The Cobham Company supplied various components of the KC-46 oxygen system.

Airlift Capabilities

The Pegasus, being a wide-body aircraft, can handle a variety of load configurations in the upper lobe:

- 18 463L cargo pallets in a twin row, or
- 10 pallets in a single row, or
- 58 FAA-certified palletized passenger seats, or
- 114 high-density palletized passenger seats, or
- 15 seats for MedEvac medical personnel, or
- 58 MedEvac patients, or

- An integrated roller system for very large, heavy cargo items.

The KC-46 can be configured into any of the above configurations within two hours.

From this rather long description of the KC-46, the reader can see that while the basic airframe design is quite old, the aircraft possesses many new technological advances, which make this tanker more valuable to the USAF and enable it to better perform its varied missions.

Unfortunately, the introduction of the KC-46 into service brought with it a number of problems, both major and minor, as we shall see in the next chapter. This range of problems delayed the acceptance of the Pegasus into operational service even further.

Conclusion

The entire early history of the KC-46 Pegasus tanker is a textbook example of government's inability to bring a major acquisition project to a satisfactory result in a timely manner. It is also a perfect example of top-level executives controlled by their greed and incompetence.

CHAPTER 23

THE KC-46 INTRODUCTION AND PROBLEMS

As explained in Chapter 21, the initial acquisition of the KC-46 got off to a rocky start with multiple delays and a Pentagon official being charged with criminal activity and being sent to jail. Then, in 2019, as the Boeing Co. began delivering KC-46 tankers to the USAF, the service was looking forward to the entry of the Pegasus into regular service. Unfortunately, that did not turn out to be quite the case.

Initial Production

The contract signed by Boeing and the USAF called for four (4) test aircraft to be built, with a follow-up delivery of 18 operational aircraft. The first Pegasus came out of the Boeing wide-body manufacturing facility in Everett, Washington, in January 2015, with the first flight of a KC-46 taking place in September 2015.

The delivery of the first four (4) aircraft manufactured meant flight testing of the KC-46 could begin. The Boeing Company completed its normal testing program, after which the U.S. Air Force began its testing of the KC-46 at Edwards AFB, California, which is where the Air Force Flight Test Center is located.

The first operational KC-46s were delivered to McConnell AFB, Kansas, in January 2019, where the process of getting them ready to fly missions began. Soon afterwards, the Air Force began indicating to

Boeing that there were a number of problems with the new tankers. These problems began to appear during the initial months of operational testing at McConnell AFB. In addition, some tankers were assigned to Altus AFB, Oklahoma, where the training of crews for the KC-46 was planned to begin. As of August 2025, a total of 89 KC-46 tankers had been delivered to the Air Force.

The KC-46 tankers are currently being delivered to the USAF at the rate of two per month.

Below is the list of problems that have plagued the KC-46 program from the start.

Problems Encountered

1. The refueling boom extending without being commanded to do so
2. The refueling boom scraping the receiving aircraft
3. Stiff refueling booms, which made them more difficult to control
4. The Refueling Visual System (RVS) lacked visual acuity, which meant the boom operator had difficulty in positioning the boom properly over the receiver aircraft receptacle.
5. Radio antennas generated electrical sparks while the aircraft transmitted.
6. Cargo locking system did not work properly.
7. Drain tubes in the refueling receptacle (for removing water) cracked at low ambient temperatures.
8. Problems with the Auxiliary Power Unit (APU).
9. The Flight Management System produced navigation anomalies.
10. Fuel tanks leaked.
11. The On-Board Inert Gas Generation System required one to two hours to complete its start-up process; takeoffs were prohibited until the system had completed this initial step.

12. Some aircraft were delivered with trash in the cabin and with fuel tank debris.

Let us now take a closer look at the most serious of these problems.

Probably the chief problem was the brand new Remote Vision System (RVS) for refueling receiver aircraft. The system, which uses three panoramic cameras, was a source of controversy from the beginning. The Air Force complained that boom operators lacked visual acuity, particularly at night or in cases of reduced external lighting. This meant that the KC-46 could not be depended on for all missions. As a result, Boeing, which had designed the system, agreed to redesign the system at its own expense. This redesign process required a further two years and was forecast to be ready in 2024.

In the meantime, Boeing proposed installing an enhanced software package that would refine the images the boom operator sees. The Air Force agreed to this step provided it did not delay the arrival and installation of the upgraded Remote Vision System. The installation of this enhancement began at Altus AFB, Oklahoma, in 2021.

A second problem was a new type of radio antenna that actually used the airframe as an antenna, caused sparks, and was potentially extremely dangerous. The antennas caused sparks during communication transmissions. If sparks occurred during refueling, and if any fuel had leaked out, fires or a catastrophic explosion could easily result.

Then there were various problems with the refueling boom, an updated boom design originally used on the KC-10 tanker. The boom remote control system was causing it to extend or retract without commands from the boom operator. The boom was also found to bump against a receiver aircraft and was considered to be stiff when receiving commands. A fix of this problem was expected to be completed by 2024.

Another problem was the Auxiliary Power Unit (APU). APUs have been in use for many decades, but in this case, it was discovered that there were two issues. One issue had to do with the drain mast, which is on the outside of the aircraft tail. It turned out this drain mast could actually break loose. Boeing checked and discovered the issue was caused by poor welding and was considered a quality control problem. The second issue had to do with the duct clamp, which vibrated and moved back and forth and could potentially fail. The fix involved a new clamp design, which was expected to solve the problem. The Air Force delayed the formal acceptance of the first batch of KC-46 tankers because of these problems. The aircraft were eventually accepted by the Air Force as capable of flight but not of performing operational missions. The cumulative result of all these problems was that the KC-46 was years late in reaching operational status. Additionally, the Air Force withheld a portion of the scheduled progress payments to Boeing until the various problems were completely fixed.

Because the contract originally signed by Boeing was of the fixed-price type, Boeing had no choice but to attempt to fix all of these problems at its own expense. Between 2014 and 2019, the net cost to Boeing to rectify these problems totaled $5.037 billion.

As of 2022, the Air Force still did not consider the KC-46 to be fully mission qualified. Furthermore, the Air Force advised Boeing that it would accept only two KC-46 tankers per month, even though the number of completed tankers parked on the Everett, Washington, plant ramp kept increasing. The accumulated delays caused the Air Force to have to make a number of changes as listed below.

1. The planned retirement of some KC-135 tankers was delayed considerably.
2. The scheduled transfer of trained personnel from the KC-135 and KC-10 units to KC-46 squadrons was also delayed.
3. Training of KC-46 flight crews and maintenance personnel was slowed.
4. Support activities meant specifically for the KC-46 were all rescheduled.

5. Note that the maintenance of the KC-135 fleet was estimated at $10.3 million per aircraft per year. This explains why the Air Force was eager to retire as much of the KC-135 fleet as feasible and replace it with the new KC-46.

A review of the problems identified with the KC-46 clearly indicates that Boeing and/or the Air Force made mistakes in the design and development of the Pegasus. Moving the boom operator's position has been questioned, along with the addition of a second boom operator's workstation. If the boom operator's position had always been left where it previously was, the basic cost of the KC-46 would have been reduced considerably, and some of the major problems would have been avoided.

An interesting result of the initial delivery to McConnell AFB came about during routine maintenance. The maintenance personnel were tasked with cleaning the belly fuel tanks located in the lower lobe under the main floor. This is not an easy job, but it is very important. Personnel discovered that Boeing had found a way to remove the belly fuel tanks so that they would be much more accessible for cleaning outside of the aircraft. This seemed an excellent solution to the problem, since working on the fuel tanks in the limited space of the belly is a difficult and lengthy endeavor. The removal, cleaning, and reinstalling of these fuel tanks was expected to require 3-5 days. The McConnell personnel did it in just 2 days per aircraft. This step should provide valuable experience for KC-46 maintenance crews.

In July 2022, the Air Force revealed that the fleet of delivered KC-46 tankers was at last being used operationally. This was in spite of the fact that the fixes for some of the problems mentioned were still months or years away from being completed.

During this initial period, when the Pegasus was used on actual missions, it racked up some noteworthy statistics. Between July 2021 and July 2022, the KC-46 fleet flew more than 3,000 operational missions. During that period, the KC-46s off-loaded 66 million pounds of fuel in more than 16,000 contacts with the refueling boom and 1,000 contacts using the drogue. This would seem to indicate that, in spite of the many problems still affecting the Pegasus, the tanker was nevertheless capable of carrying out actual missions successfully.

During that same period, the Pegasus also refueled aircraft from seven foreign countries, most of which are members of NATO. The KC-46s flew 81 missions for a total of over 536 hours, and off-loaded 2.5 million pounds of fuel to these 155 receivers.

The Air Mobility Command reported that, in spite of the existing problems with the aircraft, crews were becoming much more comfortable flying and operating the airplane. With the KC-46 being considered at least partially mission qualified, crews were able to refuel various aircraft types. This formally made those receiver aircraft operational receivers when refueling from KC-46 tankers. The Air Force also reorganized the operational KC-46 squadrons so that they would have two flight crews for each KC-46.

In September 2022, the Air Mobility Command (AMC) suddenly announced that the KC-46 had been cleared for worldwide deployments and combat tasking. Considering how critical the AMC had been of the Remote Visual System (RVS) installed in the KC-46, this announcement probably surprised quite a few people. One possible explanation for this sudden acceptance of the RVS, even though the fix is temporary, is that crews began to be more comfortable and proficient operating the system. In the meantime, Boeing is still expected to deliver RVS 2.0 by 2024, which will address some of the issues that AMC has detailed since the introduction of the Pegasus.

Conclusion

The KC-46 was finally considered ready for some, if not all, of the missions it was originally designed to carry out. In the next chapter, the very basic function of aerial refueling is described.

CHAPTER 24
HOW REFUELING IS DONE

Aerial refueling requires the successful completion of two critical steps. The first such step is for the tanker and receiver to make a *rendezvous*, which brings the two aircraft in sight of each other. The second step requires that the receiver connect to the tanker, thus beginning the transfer of fuel. Both steps must be successfully completed whether by day or by night and in various weather conditions. If refueling is done during wartime, then the entire operation may become somewhat more difficult, particularly if the entire two-step process is done without any radio transmissions so as not to betray the exact position of the two aircraft. Aerial refueling requires a considerable level of skill to be completed successfully.

Making the Rendezvous

There are two primary ways to make a rendezvous.

1. Rendezvous Delta. In this type of rendezvous, both the tanker and the receiver are given what is referred to as an *anchor point.* This is a specific geographical point defined by latitude and longitude. In some cases, the anchor point may also be defined as a point that is a certain distance and bearing from a TACAN (Tactical Air Navigation System) facility. TACAN provides distance and bearing from a TACAN ground facility by sending a

continuous signal on a specific Ultra High Frequency (UHF). TACAN is used only by the military. TACAN facilities are often housed within FAA navigational facilities known as Very High Frequency Omnidirectional Range (VOR) buildings. These are shown on aviation charts simply as a VOR. When such a facility is shared by VOR and TACAN equipment, it is identified as a VORTAC facility. USAF aircraft all carry TACAN equipment, and a tanker broadcasting on its TACAN UHF frequency, then acts as a point of rendezvous for a receiver approaching that tanker.

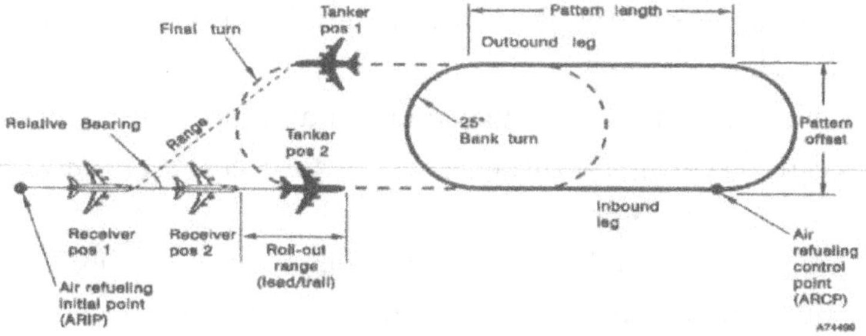

Fig. 24-1 Diagram of Rendezvous Delta

2. The tanker flies to that predetermined anchor point and establishes a left-hand racetrack pattern at a prearranged altitude. The racetrack pattern generally has legs separated by a minimum of 20 nautical miles and legs that are a minimum of 50 nautical miles long. This racetrack pattern can vary in size depending on circumstances. Once the receiver aircraft reaches the rendezvous area, it maneuvers to place itself behind and below the tanker. (See Fig. 24-1).

3. Rendezvous Golf – This is a rendezvous procedure for en route refueling. Rendezvous Golf is used when the tanker and receiver are en route to an air refueling track, anchor point, or final destination. Rendezvous Delta procedures are normally used when

refueling aircraft on transoceanic flights, whether it's a mission or a ferry flight.

4. For Rendezvous Delta, the tanker flies a specific course at a specific altitude. The receiver aircraft will be flying on an opposite course. When the receiver sees the tanker, it executes a 180-degree turn and positions itself behind and below the tanker.

In either procedure, the Boom Operator coaches the receiver into position by radio and operates the boom so that it connects with the receiver. When the prescribed amount of fuel has been transferred, the Boom Operator warns the receiver that the boom is about to disconnect.

Radio transmissions between tanker and receiver in wartime conditions can use very low-power communication gear that cannot be picked up by enemy equipment.

Refueling can be done at any altitude. Over the CONUS, it is often done at altitudes between 15,000 feet and 25,000 feet. Thus, at these altitudes, refueling is accomplished above practically all general aviation aircraft but below airline, military, and business traffic. During wartime missions, refueling has been accomplished at altitudes as low as 1,000 feet above ground in order to avoid being targeted by enemy radar and missiles or drones.

Tankers are equipped with a multitude of lights that, particularly for night refueling, can greatly assist the receiver pilot to reach the correct position for taking on fuel.

These lights are shown in the next three illustrations.

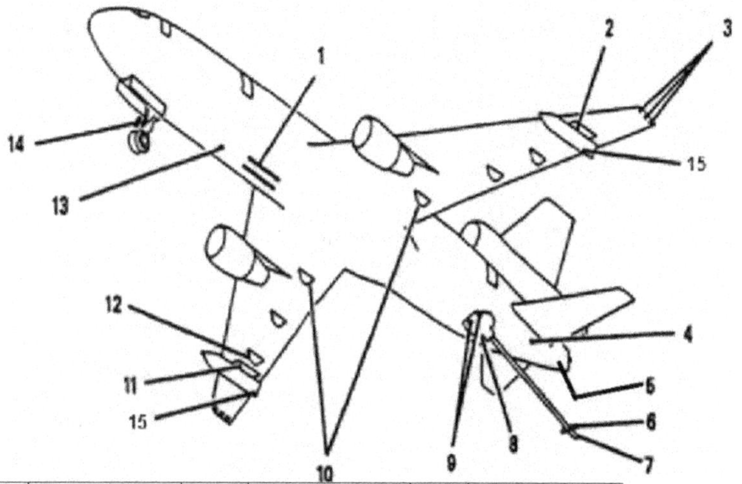

1	Director Lights	6	Boom Marker Lights (Fluorescent)	11	Inboard Pod Markings Illumination Light ***
2	Outboard Pod Markings Illumination Light ***	7	Boom Nozzle Lights	12	Aft Fuselage Underbody Light (Typical)
3	High Intensity Supplemental Lights (Typical)	8	Hose/Drogue Signal Lights	13	Lower Anti-Collision Rendezvous Light
4	Horizontal Stabilizer Illumination Light (Typical)	9	Wing Illumination Lights	14	Nose Landing Gear Lights
5	Receiver Aircraft Floodlights - in Tanker tailcone	10	Forward Fuselage Underbody Lights	15	Wing Pod Signal Lights

Fig. 24-2 Fuselage Lights

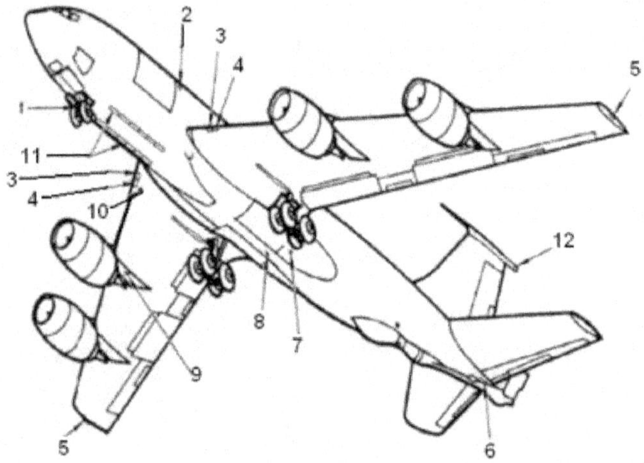

1	Nose Landing and Taxi Light	5	Navigation Light	9	Underbody Illumination (Typical)
2	Nacelle Illumination (Typical)	6	KT Boom Searchlight (Q Tankers Only)	10	Terrain Light (Retractable)
3	Taxi Light	7	Under wing Illumination Light (Typical)	11	Receiver Pilot Director Lights-for boom operator direction to receiver pilot
4	Landing Light (Fixed)	8	Strobe Lights	12	Tail Mounted Floodlight-for Boom Operator visibility of receiver aircraft

Fig. 24-3 Fuselage Lights

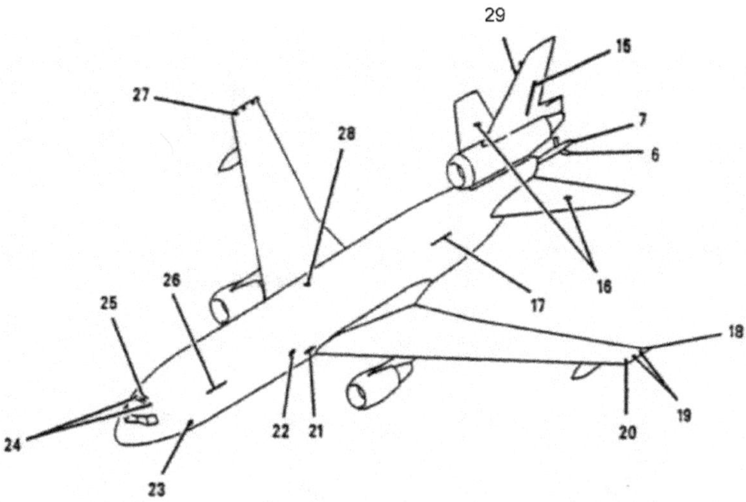

6	Boom Marker Lights (Fluorescent)	19	Formation Light (Typical)	25	Lead-In Stripes/UARRSI Electroluminescent Lighting System
7	Boom Nozzle Lights	20	Forward Position Light, Red	26	Formation Light (Typical)
15	Formation Light (Typical)	21	Taxi and Runway Turnoff Light - Ground Flood (Typical)	27	Forward Position Light, Green
16	Logo Lights	22	Wing and Engine Scan Light (Typical)	28	Upper Anti-Collision/Rendezvous Light
17	Formation Light (Typical)	23	Fuselage Landing Light (Typical)	29	Upper Fuselage Floodlights
18	Aft Position Light White (Typical)	24	Slipway Flood Lights		

Fig. 24-4 Fuselage Lights

Taking On Fuel

Once the tanker and the receiver are in sight of each other and positioned properly, the receiver has to approach the tanker from behind and slightly below. This is the most difficult part of the refueling operation. The receiver must not have an approach speed that is much higher than the tanker's cruise speed. Overtaking the tanker too quickly can and has resulted in collisions with often-fatal results.

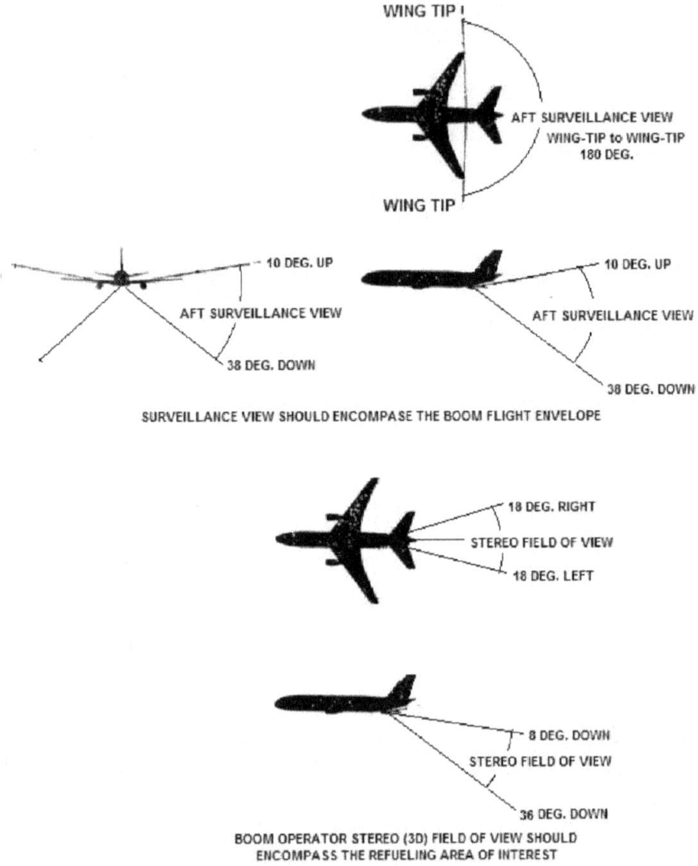

Fig. 24-6 Diagram of Boom Operator's View Limits

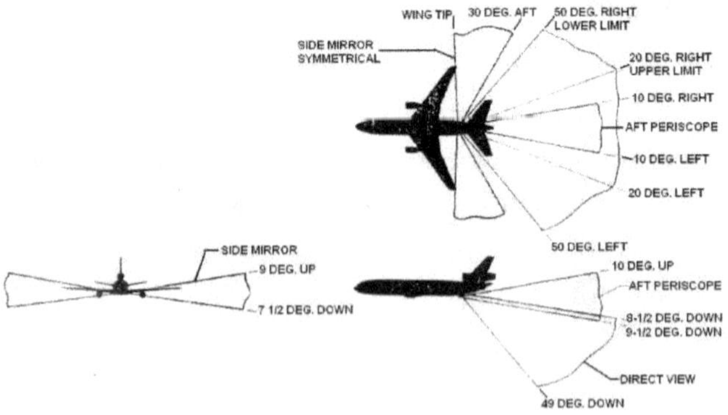

Fig. 24-7 Diagram of Boom Operator's View Limits

The tanker can then initiate refueling. If refueling a large aircraft, the Boom Operator lowers the boom, extends it, and then maneuvers it so that its end plugs into the receiver's fueling receptacle. Generally, large military aircraft have fuel receptacles located just behind the flight deck on the top of the fuselage, or in some cases, in the very nose of the aircraft. Once the boom is firmly attached to the receptacle, actual refueling will begin automatically.

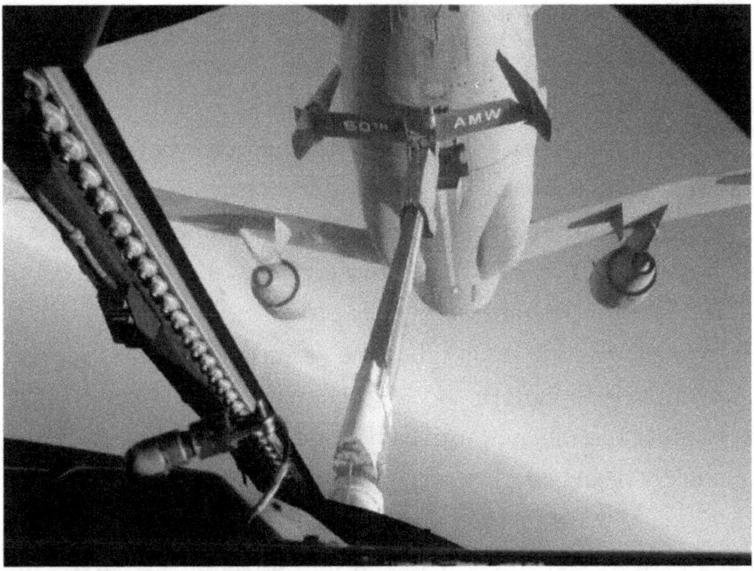

Fig. 24-8. Photo Showing Boom Operator's View of Receiver During Refueling

As fuel flows into the receiver, the receiver pilot has to compensate for the constantly increasing weight by very slowly increasing engine thrust. The receiver must exercise care to maintain the correct position so that neither the boom nor the receiving aircraft is damaged. Moreover, if the receiver aircraft moves too far beyond the correct position, the boom will automatically disconnect.

Refueling requires time. The table below shows how much time is required depending on the amount of fuel being transferred.

FUEL WEIGHT / Number of (POUNDS)	MINUTES
450,000	50
350,000	40
250,000	30
100,000	15
50,000	10
25,000	7
10,000	4
3,900	2.25

From the table above, the reader can see how transferring large quantities of fuel can require rather long periods of time. During these periods, both the Boom Operator and the receiver's flight crew are under considerable stress. The tanker flight crew has the relatively simple task of flying straight and level at a constant speed. The Boom Operator has to keep track of the amount of fuel transferred. Furthermore, the Boom Operator, who has direct radio contact with the receiver, has to watch carefully that the receiver does not drift to either side, nor move too high or too low relative to the tanker. This is critical, because if the receiver moves beyond certain limits, the boom will automatically disengage.

If refueling takes place at night, both the rendezvous and the actual refueling process become more risky and difficult. Another complication is encountering air turbulence. Depending on the severity of the turbulence, the tanker aircraft may elect to move the operation to a different altitude in the hope of finding much smoother air.

The reader needs to keep in mind that large aircraft, such as bombers and airlifters, are refueled by using the boom, while fighters refuel using the hose-and-drogue system. In the case of the drogue, the Boom Operator simply releases the drogue so that it trails the tanker's wing by some 75 feet. The fighter then approaches the tanker and is positioned close to the trailing drogue.

The receiver pilot then moves his aircraft slowly forward until the *probe* connects with the drogue. The probe is a short length of rigid pipe that protrudes from the front of the receiver and has a receptacle at its front end.

There is often confusion as to how many aircraft a tanker can refuel simultaneously. To clarify this point, the reader needs to remember certain things.

1. Tankers have a centerline boom generally used to refuel large aircraft such as bombers, airlifters, transports, and even other tankers.
2. Some tankers also have a centerline hose and drogue that can be extended below the belly of the tanker and is used to refuel fighters.
3. In addition, some tankers are equipped with removable hose and drogue pods attached at each wing tip. When these are available, two fighters can be refueled at the same time.
4. Some tankers have three refueling options. One option is to refuel a large aircraft with the centerline boom. A second option is to refuel a fighter with a centerline hose and drogue. A third option is to refuel two fighters using the two hose and drogue pods on the wing tips. The tanker cannot refuel a large aircraft using a centerline system and simultaneously refuel other aircraft using the wing-tip systems. The reason is simply that it would place receiver aircraft too close to each other.

Refueling Tracks

There are many refueling tracks all over the Continental U.S. (CONUS). These are used mainly for training purposes, to maintain proficiency, and as needed when ferrying aircraft for operational missions. Refueling tracks are assigned individual radio frequencies that are used by tankers and receivers as they maneuver to make contact. There are so many refueling tracks in the airspace over the CONUS that showing an illustration of their location would be pointless.

Planning A Refueling Mission

Normally, refueling missions are planned well ahead of time.

A number of specific items must be accomplished correctly as part of the planning, as listed below.

1. Location of rendezvous.
2. Whether the rendezvous is to be in daytime or at night.
3. Making the rendezvous at night or in poor visibility can be an issue. Generally, one mile of visibility is necessary for the receiver aircraft to visually pick up the tanker. A night mission makes the initial contact much more difficult. Occasionally, if visibility is simply too restricted by cloud cover, the rendezvous may have to be changed to a different altitude. Another factor is that some receiver aircraft may not have deicing equipment, and therefore the tanker must ensure that the refueling is carried out where there are no icing conditions. An occasional but rare factor that may occur when making a rendezvous is the presence of cloud layers in the distance, which can cause illusions and disorientation.
4. Refueling fighters using the hose and drogue is more difficult because the drogue may move around due to air currents. Thus, fighter pilots must be quite skillful to successfully connect their refueling probe with the drogue. It takes practice to accomplish this consistently.

Under normal conditions, the receiver aircraft will contact the tanker while up to 40 miles distant. At night or in limited visibility, radar in both aircraft is used to accomplish the rendezvous. Radar is also used when flying in wartime conditions and radio communications are forbidden. At night, special lights on the rear of the tankers greatly help the receiver aircraft reach the correct position.

Above all, tanker crews are keenly aware of the vital importance of completing the missions satisfactorily. In previous chapters, some examples were described when tanker crews managed to complete their assigned missions under less than favorable conditions. Furthermore, tanker crews are always alert to emergency radio calls from aircraft that need fuel, even if not scheduled to refuel a particular receiver. Flexibility is a critical component of a tanker crew's thinking.

On long-distance flights, especially for fighters flying over open water, a tanker may accompany a fighter aircraft across the Atlantic or Pacific Oceans, refueling it frequently in case the receiver has to recover at an alternate airport because of an emergency.

In an interview with Col. Mark E. Smith, USAF (Ret.), a former fighter pilot, I asked him whether he found refueling from a tanker particularly difficult. His response was that the first few hook-ups could be somewhat intimidating. But after doing it a number of times, it became quite routine. Col. Smith also pointed out that as a fighter pilot, he had to become proficient in formation flying, and this helped considerably in making the correct approach to a tanker and then completing the hook-up.

Flying a Refueling Mission

In the summer of 1994, this writer was able to experience actual refueling missions. As an officer in the Civil Air Patrol (CAP), I had been tasked to escort a small group of CAP cadets to the annual air show in Oshkosh, Wisconsin. The transportation was provided by the Illinois National Guard 126th Refueling Wing, at that time based at Chicago O'Hare International Airport. The tanker was a KC-135. We departed Chicago O'Hare and climbed to an intermediate altitude over Lake Michigan. Somewhere north of Chicago, we made a rendezvous with a flight of four A-10 fighter-bombers based at Battle Creek, Michigan. The tanker refueled all four on the way to Oshkosh and an additional four on the return flight.

Conclusion

The air refueling of aircraft is not a simple exercise. But with the proper training, it now is done frequently, successfully, and above all, safely.

The total set of responsibilities of the tanker fleet is actually somewhat complex, and in the next chapter, these are listed and described.

CHAPTER 25
OVERALL REFUELING RESPONSIBILITIES

Air refueling (AR) operations have come a long way since the early days, when KB-29 tankers first began developing the basic concept. Initially, AR was seen as a means of extending the range of nuclear bombers, such as the Boeing B-47. Since that time, AR has expanded and matured. This chapter will help explain the many missions of aerial refueling.

Refueling Airspace

AR has become so prevalent that the USAF and the FAA realized that for both safety and operational efficiency, certain areas of airspace had to be set aside for the specific use of AR within the Continental United States (CONUS). The airspace that is set aside for AR is used in two ways: tracks and anchors. (See Chap. 24). These refueling areas have specific boundaries as well as refueling altitudes. In addition, tracks to and from the refueling areas are spelled out. This information is published and is available to all aircraft flying within the CONUS. The FAA Air Route Traffic Control Centers (ARTCC) use this information to ensure air safety.

Refueling within CONUS generally takes place between 15,000 and 25,000 feet. The lower limit is above the great majority of general aviation aircraft, while the upper limit is below commercial, business, and military traffic.

Nuclear Operations Support

AR of USAF bombers was the original mission of air tankers. It still ranks as a primary mission. Much of this mission falls under the category of training within the CONUS. However, during actual bomber missions designed to show up on enemy radar, much of the AR will take place over water.

Reconnaissance Support of Nuclear Missions

These types of missions by reconnaissance aircraft are in support of long-range bomber missions. Tankers, in their turn, support the reconnaissance aircraft. Supporting the reconnaissance assets enables them to provide enhanced and vital information to the nuclear strike force.

Command and Control Aircraft Support

The USAF Command and Airborne Warning and Control (AWACS) aircraft often must remain airborne for very long periods. AR makes it possible for these types of aircraft to remain aloft while far from any available airfield.

Global Strike Support

This type of support has repeatedly been shown to be truly invaluable. In previous chapters in this book, the ability of AR to enable both air assets and their supporting elements to operate in distant areas has been clearly demonstrated.

Air Bridge Support

Air bridge support refers to the AR that enables USAF assets to reach distant areas beyond the CONUS without time-consuming fuel stops. Basically, this involves en-route fueling. It also makes possible air bridges when other nations do not allow U.S. aircraft to either land in or fly over their airspace.

Aircraft Deployment Support

This type of support is similar to Air Bridge Support in that it involves the deployment of assets to points beyond the CONUS. It may differ somewhat in that the deployment may be intermittent.

Theater Support (Combat Air Refueling Support)

This type of support may be critical since it enables combat aircraft, which may be based well away from the combat zone, to reach their assigned targets and still return safely to their bases. In addition, these tankers are available to assist combat aircraft that are either damaged by enemy action or have a malfunction.

Special Operations Air Refueling Support (SOAR)

This type of support can often take place over enemy territory and at night. In many instances, AR is critical if the mission is to be successfully completed.

Joint Multinational Support

The U.S. fleet of tankers has the ability to refuel aircraft from a number of foreign countries. During wartime operations, when coalitions with some of these countries may be formed, multinational support becomes extremely important.

Training Support

The tanker fleet has a continuing responsibility to train not only its own tanker crews but also to be available to train aircraft crews of the U.S. Air Force, Navy, and Marines. This support is visible in the large number of AR tracks in use in the CONUS.

Conclusion

The total set of responsibilities of the tanker fleet is truly awesome. In the next chapter, the gradual increase of the capabilities of the KC-46 tankers is explained.

CHAPTER 26
KC-46 OPERATIONS

When deliveries of the KC-46 Pegasus began in January 2019, some of the tankers went to Altus AFB and some to McConnell AFB. Altus AFB is located in southwest Oklahoma, and the initial training of KC-46 crews was carried out there. Other Pegasus tankers went to McConnell AFB near Wichita, Kansas, where the initial operations began. These initial operations involved routine missions refueling various USAF aircraft. By flying these missions, the 22nd Air Refueling Wing, located at McConnell AFB, was checking to see how compatible each type of aircraft in the Air Force inventory was when being refueled by a KC-46 tanker. These missions also helped sharpen the skills of the KC-46 refueling operators using the new remote systems. In May 2022, the Air Mobility Command (AMC) decided to see how the Pegasus would perform on a truly long-range mission. On May 5, a KC-46 of the 22nd Air Refueling Wing departed McConnell AFB. The flight had been planned to last 24 hours. To accomplish this, the aircraft was crewed by a total of 11 personnel. There were six pilots, three boom operators, a photojournalist, and a physician's assistant. Two pilots were on duty for four hours before turning the flying duties over to the next two pilots.

The flight path was a large figure eight over the CONUS. The track began at McConnell AFB in Kansas, went north and then northeast to South Dakota, north to the U.S.-Canadian border in Minnesota, southeast to Chicago, Illinois, northeast to Maine, south along the Atlantic

Coast to Key West, Florida, northwest back to McConnell AFB, west to Los Angeles, north to the state of Washington, and finally southeast back to McConnell AFB. That route covered 9,000 nautical miles and required 24.2 hours to complete. The flight ended where it departed, McConnell AFB.

During the long flight, the KC-46 made a dry contact with another KC-46, refueled four Marine Corps F-35 fighters, and was in turn refueled by another KC-46. Cots were set up in the cargo bay so that off-duty crew members could get a few hours of sleep.

By comparison, in December 1992, a KC-135R of the Air Mobility Command flew from Kadena AB, Okinawa, to McGuire AFB in New Jersey, non-stop and without refueling in 17 hours 31 minutes.

Unfortunately, despite the increased operations, the KC-46 continued to have a variety of problems. One example occurred in August 2022. A KC-46 of the New Hampshire Air National Guard based at the Pease ANG Base flew a refueling mission while carrying a group of 16 staff employees who work for the New Hampshire Congressional delegation. The flight was a demonstration of air refueling and of the capabilities of the KC-46 Pegasus.

At the end of the refueling, the crew tried to retract the refueling boom. The cable that retracts the boom snapped, leaving the boom hanging down. The Aircraft Commander had to divert the flight to Joint Base McGuire-Dix-Lakehurst, New Jersey, because the runway at Pease AFB was closed when a USAF C-5 Galaxy transport landed there due to an unspecified emergency.

The KC-46 landed safely at McGuire with no injuries. Damage to the KC-46 is unknown.

In November 2022, it was revealed that even as brand-new KC-46 tankers are entering the Air Mobility Command, some of the previously delivered KC-46s were being taken out of service for scheduled overhauls. These included the changes to the Remote Vision System, a system that Boeing had promised to change by 2024. Other overhauls included work to be done on the drain masts, the Auxiliary Power Units, and a fix for the booms.

All of this work required lengthy periods during which each KC-46 was out of service. This work may also result in a further delay to the retirement of KC-135 and KC-10 tankers.

The gradual increase in missions flown by KC-46 tankers was slower than expected, but it did increase as more were delivered to the USAF.

Conclusion

During this process, the Pegasus proved to be a formidable tanker and airlifter. In the meantime, the tanker fleet was preparing to undergo further changes in its make-up and operations.

CHAPTER 27
FOREIGN TANKERS

Over the years, a number of foreign countries acquired aerial tankers. Many of these were aircraft from the European aerospace consortium known as Airbus. The reader should note that the information shown below is likely to be incomplete.

Avro Lancaster

The Avro Lancaster was a World War II reciprocating four-engine heavy bomber operated by Great Britain's Royal Air Force (RAF). In 1948, the RAF acquired four of these aircraft converted to a tanker role and equipped with the probe and drogue system. These were later retired.

Avro Lincoln

The Avro Lincoln was an improved version of the Lancaster and was also equipped with a probe and drogue system. An unknown number were built for the RAF as tankers and later retired.

Avro Vulcan

The Avro Vulcan was originally a four-engine jet bomber. Six of these were equipped with a probe and drogue system for the RAF in 1982 and later retired.

Boeing KC-135

A number of KC-135 tankers were sold to foreign countries. Among these were Chile, France, Singapore, and Turkey.

Vickers VC10

In the early 1960s, the RAF decided to acquire some jet aerial tankers. These were Vickers VC10 commercial jet aircraft that were then converted. These aircraft were a combination of the standard VC10 and the more powerful engines and tail-fin fuel tank of the Super VC10. These aircraft also had a detachable nose probe for refueling and an APU in the tail cone. A total of fourteen of these combination aircraft were delivered to the RAF in the mid-1960s.

These tankers were powered by four Rolls-Royce Conway turbofan engines of 22,500 lbs. of thrust each, which resulted in a maximum take-off weight of 334,878 lbs. These VC10s had a maximum fuel load of 170,000 lbs. The RAF used them on two principal routes: one to the Far East and Singapore, and the other to New York. By 1970, these aircraft were carrying 10,000 passengers and 730,000 pounds of cargo monthly.

In the late 1970s, an additional five former commercial VC10 and Super VC10 aircraft were converted to tankers for the RAF.

Airbus A310

The Airbus A310 MRTT (Multi-Role Tanker Transport) was a conversion of the A310 commercial jet transport. It had a maximum take-off weight of 361,558 lbs., a maximum speed of Mach 0.79, and could offload 88,000 lbs. of fuel, or carry 79,000 lbs. of cargo, or carry 214 passengers. The A310 was powered by two GE or Pratt & Whitney turbofan engines with a thrust of 79,000 lbs. each. The aircraft was equipped with three hose-and-drogue refueling systems. Four of these tankers were delivered to the German Air Force beginning in 2004. In addition, the same year, six of these aircraft were sold to Canada.

Airbus A330

The Airbus A330 MRTT is a twin-engine, wide-body tanker that is a variant of the commercial A330 jet transport. It has a maximum take-off weight of 466,000 lbs. The A330 MRTT had a total of 55 orders from many different countries, as well as NATO.

The A330 was a wide-body aircraft originally produced as a passenger transport. The MRTT version was powered by two (2) turbofan 72,000-pound thrust engines, which could be Rolls-Royce, GE, or Pratt & Whitney. The aircraft had a maximum take-off weight of 514,000 lbs. and could carry up to 245,000 lbs. of fuel.

In 2011, seven (7) of the A330 MRTT tankers were delivered to the UK for the Royal Air Force.

The A330 MRTT was sold to various countries, with a total of 61 aircraft delivered. Countries receiving these aircraft included Australia, Saudi Arabia, Spain, and the United Arab Emirates.

IL-78

The Ilyushin IL-78 is a Russian aerial tanker that first flew in 1983. It was based on the IL-76 wide-body transport. It has a maximum take-off weight of 472,000 lbs. and can transfer up to 220,000 lbs. of fuel. Fifty-three (53) were built for the Russian Air Force.

H6-U

The H6-U tanker is a Chinese conversion of Russia's Tu-16 Badger medium bomber. China is believed to have 24 tankers. They have a limited range and can offload approximately 37,000 lbs. of fuel. Compare this with the 180,000 lbs. of fuel carried by the Boeing KC-135 operated by the U.S. Air Force's Air Mobility Command.

YU-20

The Y-20 is China's new tanker currently under development. It is a conversion of the Y-20 military transport, which is reportedly of similar size to the Boeing C-17 military airlifter of the U.S. Air Force.

As of 2025, it is clear that the overwhelming majority of air tankers in the world belong to the U.S. military.

CHAPTER 28
THE TANKER FLEET UNDERGOES CHANGE

KC-135 Retirements

Beginning in October 2003, the USAF initiated a slow retirement of its KC-135 Stratotankers, some of which were 46 years old by that time. One of the reasons for these retirements was the fact that maintenance of the KC-135s was becoming ever more expensive.

KC-10 Retirements

On July 13, 2020, the first KC-10 Extender was retired. It was flown to the Davis-Monthan AFB storage facility in Arizona. The KC-10 retirements are planned to stretch over a period of five years.

The retirement of older tankers is predicated on the arrival of new KC-46 tankers, a program which is running years behind the original schedule. As a result, the actual retirement of KC-135 and KC-10 tankers will be at a slower pace than originally planned.

As of September 2025, the USAF still retained a total of 398 KC-135s as the backbone of the tanker fleet. The retirement date of the last of these aging tankers is said to be in 2050. But that date could easily change. One factor that might change that date is the increasingly high cost of maintaining those KC-135s. A second factor is the grounding of KC-135s due to crashes caused by age. A third factor could eventually be that a Federal Administration, together with Congress, decides to fund either many more KC-46 tankers or possibly even a tanker of totally new design. The fact is that by the year 2050, the KC-135 tanker fleet will be close to 90 years old. These aircraft were never expected to be still flying for so many years. Still another factor is how international relations develop in the next few years. In June 2025, a military operation took place that was another powerful reminder of the importance of the USAF tanker fleet.

Operation Night Hammer - June 21-22, 2025

This operation was authorized by the Federal Government because Iran was secretly continuing its development of nuclear weapons despite a treaty which forbade that and to which it had agreed.

In response, the USAF planned a raid called "Night Hammer" to neutralize the Iranian nuclear facilities. The operation involved seven B-2 Stealth bombers as the primary striking force, plus a number of tankers. In addition, an unknown number of other B-2 bombers, possibly seven more, flew to Andersen AFB, Guam, in the Marianas Islands. This was a decoy force which must have used tankers to make it non-stop to Guam. It is not known which tanker models were employed in this operation. Also participating were a number of escorting fighters launched from Germany, plus a total of approximately thirty tankers. In all, over 150 aircraft were reported as having participated in the operation, which was considered a total success.

Without the use of the tanker force, it probably would have been extremely difficult, if not downright impossible, to mount Operation Night Hammer. Once again, the tankers proved their value.

Conclusion

The tankers participating in Operation Night Hammer made that operation possible and may even have made it look easy to some onlookers.

However, one constant remains in place: the ever-present danger associated with air tanker operations. This is not merely the danger that relates to jet fuel and the possibility of collisions during refueling operations. Air refueling crews understand all too well how critical their missions really are. That awareness translates to crews being willing to fly in weather that might otherwise ground other military flights. The existence of these dangers can be seen in the lists of tanker incidents, accidents, and crashes. These are shown in the accompanying appendices.

Beyond that, when tankers operate over enemy territory there is the ever-present danger of enemy ground fire, fighters or _____. Tanker crews understand that all too well.

CHAPTER 29

THE WAY FORWARD

Based on the history of aerial refueling, it should be obvious to the reader that aerial refueling is not merely important but that its importance has steadily grown. One merely has to review the U.S. military operations from the late 1940s forward to realize that many, if not all, of these operations would have been impossible without air tankers. Simply put, the USAF air tanker fleet has been the single military unit that has enabled the U.S. to project its military force and political decisions anywhere in the world they may be required.

By the summer of 2022, the tanker fleet had shrunk to 479 tankers, a number mandated by Congress. Of that number, 59 were KC-10 tankers, approximately 56 were the new KC-46A Pegasus, and the remaining 364 were KC-135R tankers. None of the KC-135 tankers are less than 55 years old. The USAF would like to retire all the KC-135 and KC-10 tankers. That means by 2025, the entire tanker fleet would consist of approximately 150 KC-46 Pegasus aircraft. But is that number of tankers sufficient if another military crisis arises? That is the critical question that faces Congress in the middle of the decade.

The USAF is currently trying to obtain permission from Congress to reduce the total number of tankers to 455. But that would still leave a large number of legacy KC-135 tankers flying. The KC-135 is getting old for an operational aircraft. There is always a danger of corrosion and metal fatigue with old aircraft, especially in the wings. Inspecting those wings and replacing components such as wing spars is a slow and very expensive process.

In addition to the problem of obsolescence, if the tanker fleet is whittled down to the 179 KC-46 tankers currently on order, that could mean a critical shortage of tankers if the U.S. becomes embroiled in another limited war. And what would happen if the U.S. had to fight two limited wars in two distant parts of the world?

There has been talk of ordering more KC-46 tankers beyond the current contract for 179. That might be a possible solution depending on how fast Boeing could deliver these aircraft.

The USAF is currently evaluating various aircraft designs for a future tanker. Listed below are some future possibilities.

1. One of these is a blended-wing concept that would look quite similar to the Northrop Grumman B-2 bomber.
2. A Boeing 777X, which is a wide-body design, could carry a significant amount of fuel and would have a large cargo capability.
3. Another Boeing aircraft that could be considered is the 787-9 wide-body transport. This aircraft uses a significant amount of composite material, resulting in a considerably lighter airframe.
4. The USAF has also indicated it might be interested in a relatively small tanker that it probably needs in large numbers.

At this time, it is not clear what a future tanker will be or when the USAF will choose a particular design.

The way forward does not seem to be clear. A multi-year acquisition strategy for a new tanker design would appear to be the next logical step.

CHAPTER 30
A LONG VIEW OF AERIAL REFUELING

It is always very difficult to take a historical view of any extended event when that event is still occurring. Nevertheless, this writer is going to attempt precisely that.

Aerial refueling, as of 2025, can be said to have reached a certain level of maturity and to be considered an indispensable weapon system of the U.S. military. Additionally, as of 2025, aerial refueling has reached 74 years of continuous operation, beginning with the KB-29 tanker and extending to the current KC-46 tanker. Along the way there were the KC-97, the KC-135, the KC-10, and now the KC-46 Pegasus The mere fact that in that space of time a grand total of 2,360 tankers, either built as tankers or modified to become tankers, is an indisputable testament to the importance of air tankers as part of the overall military force of the United States.

Looking back over that span of time and those various tanker models, certain things are clear. Aerial refueling is now a fully accepted operation that is part of the U.S. Air Force and Navy. It is carried out daily and, depending on the international state of affairs at any given time, may be required anywhere in the world. But that is only one part of the picture.

In fact, without aerial refueling many of the us military activities would have been impossible.

Aerial refueling has developed to the point that it concurrently carries out the following missions:

1. Aerial refueling of bombers, fighters, and airlifters anywhere in the world.
2. Tankers can and do supplement airlifters by moving troops and cargo, providing humanitarian aid and evacuations, and conducting medical airlifts.
3. Tankers are also employed on reconnaissance missions.

In the simplest terms, aerial refueling is what makes possible the delivery of military force to any location on Earth within a short span of time. Considering that fact, can the value of aerial refueling be overstated?

It is a proven fact that aerial refueling has enabled the U.S. to carry out its political goals anywhere in the world. Moreover, it can do so in hours and days rather than the weeks and months it would otherwise take.

The widespread use of aerial tankers has completely changed how military operations are carried out. Today, aerial refueling has become such an integral part of military operations that it can truly be said the military can no longer operate without its presence. When aerial refueling first had its humble beginnings, who could have predicted that one day it would prove to be such a critically important and indispensable part of the military?

APPENDICES

APPENDIX A: List Of KC-97 Tanker Accidents

- June 1954 – A KC-97G en route from Altus AFB, Oklahoma, to March AFB, California, diverted to Norton AFB, California, due to weather. Crashed into Box Spring Mountain, north of Riverside. All 14 occupants killed.
- May 1955 – A KC-97G crashed into the Atlantic Ocean. An engine fire resulted in loss of control. All nine occupants were killed.
- July 1956 – A KC-97E crashed into a woods 45 miles northeast of Goose Bay, Labrador, Canada, due to an engine fire. All six occupants were killed.
- January 1957 – A KC-47G from Griffiss AFB, New York, crashed into the Adirondack Mountains. All seven occupants were killed.
- May 1957 – A KC-97F en route from Sidi Slimane AB, Morocco, to Lajes, Azores, ditched into the Atlantic Ocean 344 miles SE of the Azores due to double engine failure. All seven occupants were saved. The aircraft floated for 10 days and was destroyed by U.S. Navy battleship USS Wisconsin.
- July 1957 – A KC-97G from Plattsburgh AFB, New York, exploded and crashed into Lake Champlain three minutes after take-off. Two engines failed. Five occupants were killed and three survived.
- October 1957 – A KC-97G from Walker AFB, New Mexico, on a 10-hour low-level survey flight crashed 35 miles north of Flagstaff, Arizona. The aircraft struck a cloud-shrouded mountain. All 16 occupants were killed.

- July 1959 – A KC-97 from Pease AFB, New Hampshire, crashed near Andover, New Hampshire, at night. All seven occupants were killed.
- December 1959 – A KC-97G from Westover AFB, Massachusetts, collided with a B-52 during a refueling mission. The B-52 flew too close to the tanker, rammed the flying boom, and destroyed a major part of the tail assembly. The crew was able to make a successful emergency landing at Dow AFB, Maine. All seven occupants survived. Two of the B-52 crew ejected and were recovered safely. The B-52 landed safely at Westover AFB, Massachusetts, with the remaining eight occupants.
- March 1960 – A KC-97 ditched in the Atlantic Ocean near Cape Canaveral due to hail damage that caused engine failures. Of the 14 occupants, 3 died and the rest survived.
- April 1960 – A KC-97G crashed on take-off at Lincoln AFB, Omaha, when the landing gear collapsed. All seven occupants survived.
- June 1960 – A KC-97G from Plattsburgh AFB, New York, was refueling a B-47 on a training mission when the impeller shaft on the No. 1 engine failed due to lubrication failure. The engine caught fire. The B-47 disengaged from the flying boom. The crew attempted unsuccessfully to extinguish the fire. The left wing failed near the No. 1 engine and the aircraft went into a spin and crashed near Newry, Maine. All five occupants were killed.
- June 1964 – An HC-97G and an HC-54C were on a photo mission for the NASA Gemini program, flying near Kindley AB, Hamilton, Bermuda. During the mission, the HC-54C collided with the KC-97G, shearing both halves of one wing and the tail section. Both aircraft crashed into the Atlantic Ocean. All ten occupants of the two aircraft were killed.
- November 1964 – A KC-97 from Pease AFB, New Hampshire, crashed on take-off. All five occupants were killed.

- December 1964 – A KC-97G landing at Ernest Harmon AFB, Stephenville, Newfoundland, Canada, ran off the end of the runway after touching down too far.
- December 1964 – An Air National Guard KC-97L on a training flight was making touch-and-go landings at Fort Worth-Greater Southwest Airport, Fort Worth, Texas. On the fifth such landing, the aircraft was too low, struck an embankment short of the runway, broke up, skidded onto the runway, and burst into flames. Of the 11 occupants, four were killed.
- January 1969 – An Air National Guard KC-97L crashed short of the runway at Billy Mitchell Field, Milwaukee, Wisconsin. Four of the 11 occupants were killed.
- November 1970 – An Israeli Air Force KC-97 was on a night take off at Tel Aviv Lod Airport, Israel, when the aircraft commander saw another KC-97 being towed across the runway. He attempted to take off but the left wing struck the other tanker. Both aircraft exploded and burned. Three occupants survived and two were killed.
- September 1971 – An Israeli Air Force KC-97G was shot down by an Egyptian missile over Suez, Egypt, while on a survey flight. Seven of the eight occupants were killed.

Appendix B: List Of KC-135 Accidents

The KC-135 fleet of tankers suffered a total of 65 aircraft lost in accidents between January 1968 and May 2013.

Boeing originally delivered 732 KC-135 tankers to the U.S. Air Force. That would mean that 9% of the total fleet was lost over the years. Considering the very large number of refueling missions completed over the years, especially during wartime operations, the rate of accidents is actually quite low. It is important to remember that, with two aircraft in close proximity and considering that refueling operations occur day and night, in all kinds of weather, the low accident rate testifies to the professionalism of both tanker and receiver aircraft crew members. The reader should also keep in mind that aerial refueling is always potentially dangerous since there is a minimum of at least two aircraft that must rendezvous and then fly in very close formation.

Breaking down the total number of aircraft lost by decades is quite illuminating, as shown below.

1950s – 4 (Two years only)
1960s – 41
1970s – 11
1980s – 4
1990s – 2
2000–2013 – 2

The decade of the 1960s was obviously a very bad decade, with 41 aircraft lost. This may have been partly due to the ramping up of units as new tankers were delivered and crews were being trained. In addition, the nine-year Vietnam War was likely a contributing factor.

CONTRIBUTORS

Gordon, David, Capt., USAF (Ret.)

Moore, Michael, Lt. Col., USAF (Ret.)

Busick, Marc, Col., USAF (Ret.)

Hopkins III, Robert S. *Boeing KC-135 Stratotanker*. Midland Publishing Limited, 1997.

Susher, Harry, Jacob Benscoter, Mark Burke, Dexter Kalt, and Joshua Tokarz. *Aerial Refueling Boom/Receptacle Guide.*

U.S. Air Force Document, 2017.

Photo Credits

Fig. 2-1: photo credit: Original image by W. Wanderer. Uploaded by Mark Cartwright, published on 29 April 2024.

Fig 2.2: credit: Flickr.Original image by Airwolfhound. Uploaded by Mark Cartwright, published on 2 April 2024.

Fig. 2.3: Courtesy of the Library of Congress

Fig. 3-1: Photo of Lemay, Courtesy of the Library of Congress

Fig. 4-1: Map of Korea, Courtesy of the Library of Congress

Fig. 6.2: Credit: U.S. Air Force photo by J.M. Eddins Jr. – Wikimedia Commons, Public Domain, https://commons.wikimedia.org/w/index.php?curid=108353417

Fig. 6-3: Diagram of Hose and Drogue refueling system. Aerospace: https://creativecommons.org/licenses/by/4.0/

Fig. 7-2: The Boeing B-47 Jet Bomber, Credit: https://model-copy.info/product_info.php?language=en&products_id=20862 *the product does not require mandatory licensing and certification. All files presented on the site are our own developments and developments of our partners who have granted permission for their publication.

Fig. 7-3: The Boeing B-52 Jet Heavy Bomber. Credit: https://model-copy.info/product_info.php?language=en&products_id=20862 The product does not require mandatory licensing and certification. All files presented on the site are our own developments or those of our partners who have granted permission for publication.

Fig. 8-1: Boeing C-97 Prototype side view. Wikipedia: CC BY-SA 2.0

Fig. 14-3: Map of Indochina: Credit: By Bearsmalaysia - Own work, CC BY-SA 3.0, https://commons.wikimedia.org/w/index.php?curid=66296735

ABOUT THE AUTHOR

Claude G. Luisada is a native of Italy who emigrated to the U.S. at age 7. He has been associated with aviation in one way or another since his teens. He was at one time a student pilot. In 1956 he became a member and volunteer of the U.S. Air Force Auxiliary known as the Civil Air Patrol (CAP). He flew with the CAP as a navigator/scanner and accumulated over 700 hours. Through the years he was an officer and served in both staff and command positions. Today he is still a member of CAP with the rank of Lt. Colonel.

In 1968, as a freelance writer, he contributed many articles to the first aerospace encyclopedia, *Above and Beyond.*

In the 1970s, he published a book on the history of the Lockheed Constellation family of commercial and military aircraft. He later published a book on the history of the Boeing 787 jet transport, a groundbreaking aircraft.

Mr. Luisada has lived in Boston, Chicago, and Albuquerque and now resides in Aiken, SC.

www.ingramcontent.com/pod-product-compliance
Lightning Source LLC
LaVergne TN
LVHW010720061126
851857LV00014B/1267
* 9 7 8 1 9 6 8 5 4 8 2 4 7 *